AF335544

THERMODYNAMIC THEORY OF THE EVOLUTION OF LIVING BEINGS

THERMODYNAMIC THEORY OF THE EVOLUTION OF LIVING BEINGS

G.P. GLADYSHEV

NOVA SCIENCE PUBLISHERS, INC.
COMMACK, NEW YORK

Creative Design: Gavin Aghamore
Editorial Production: Susan Boriotti
Art Director: Maria Ester Hawrys
Graphics: Frank Grucci
Manuscript Coordinator: Phyllis Gaynor
Book Production: Joanne Bennette, Michelle Keller, Ludmila Kwartiroff,
        Christine Mathosian Tammy Sauter and Tatiana Shohov
Circulation: Iyatunde Abdullah, Cathy DeGregory and Annette Hellinger

Library of Congress Cataloging-in-Publication Data
available upon request

ISBN 156072-457-2

Copyright © 1997 by   Nova Science Publishers, Inc.
        6080 Jericho Turnpike, Suite 207
        Commack, New York 11725
        Tele. 516-499-3103   Fax 516-499-3146
        E-Mail: Novascience@earthlink.net

Printed in the United States of America

CONTENTS

FOREWORD

In this work we present a theory, rather consistent, to our opinion, which enables one to explain the origin and evolution of living beings from the view point of thermodynamics. The strong belief that the basic principles of natural science and precise theories are stable made it possible to create during the recent years the thermodynamic model of ontogenesis, philogenesis, and the biological evolution in general. The theory was met by some colleagues with understanding. To a considerable extent, this theory was developed owing to the fact that the author did not 'stick' to fashionable view points and tendencies, which 'make the absolute' out of the thermodynamics of non-equilibrium processes.

To be sure, new possibilities will appear in future for the development and application of the thermodynamic theory. New models, and more advanced ones, will be suggested. Still the author is convinced that they will be also based on the fundamentals of classical natural science.

Now, two decades after the first publication on the thermodynamics of biological evolution, it is worth mentioning that this work saw the light due to the personal decision of the late Dr. James Danielli, F.R.S., editor in chief of the Journal of Theoretical Biology. Dr. Danielli - the author of the first model of a biological membrane, an outstanding theoretical biologist and physical chemist of our century, - decided to forward the manuscript of the paper (J.Theor.Biol. 1978. **75**. P.425-444) to eight reviewers. As a result of such a thorough reviewing, the editor in chief sent a letter to the author. In particular, it ran: "Today I have decided to publish your work, in spite of the fact that Western referees faced serious difficulties while reviewing it. The reason for my decision is that this work can turn out to be quite outstanding..." A couple of years later J.Danielli wrote to the author: «I am very glad to give any help which I am able to do insisting that your ideas are appropriately published». The correspondence was continued. Unfortunately, many letters perished as time passed. However, they left a fadeless memory in my heart and always inspired me to make my way through the 'murky wood' of the extremely complicated but interesting problem of understanding life as a phenomenon. At present, the 'wood' became much thinner, there is clear sky in sight with the bright 'star of thermodynamics', which never let anybody down.

Another great scientist - Kenneth G. Denbigh, F.R.S., - discovered the 'rational grounds' already in the early publications of the author devoted to this problem. The outstanding, excellent works by K.Denbigh in various fields of natural science, as well as his personal letters, stimulated the author in his attempts to achieve the results in their most accomplished form.

The research carried out by the author in the field of biological evolution was supported by the attention of several creators of modern natural science, such as H.Urey, P.Mitchell, S.Ponnamperuma, N.Bogolyubov, Ya.Zel'dovich.

A lot of kind suggestions was made by G.Arrhenius, V.Kazakov, F.Komarov, E.Ovcharenko, Lord G. Porter and Z.Sabirov. The author is greatly indebted to all of them.

The length of the monograph is deliberately limited; it deals only with the evolution of structures considered in the framework of the thermostatic approach. The text contains some clarifying comments, and in the Appendices several notions are defined. All this is done to make the material more comprehensible.

I hope that the present work will meet understanding in the scientific world and be of interest, first of all, for young scientists.

Georgi Gladyshev

Professor of Physical Chemistry

(Institute of Ecological Biophysical Chemistry of the Academy of Creative Endeavours,

Laboratory of Thermodynamics and Macrokinetics of Non-equilibrium Systems of N.N.Semionov Institute of Chemical Physics of the Russian Academy of Sciences),

Professor of the Chair of Biomedical Systems and Devices

(N.E.Bauman Moscow State Technical University, 1975-1985)

*"Die Energie der Welt ist constant.
Die Entropie der Welt strebt einem Maximum zu."*[1]
>   **R.Clausius & J.W.Gibbs**

*"One of the principal objects of theoretical research
in any department of knowledge is to find the point of
view from which the subject appears in its greatest
simplicity."*
>   **J.Willard Gibbs**

*"...The properties of living things are the outcome of
their physical and chemical composition and
configuration."*
>   **Thomas Hunt Morgan**

*"In addition to entropy there may well exist other
'one-way' functions which add to the overall
description of the world as temporal development."*
>   **Kenneth G. Denbigh**

*"The origin and evolution of life is the origin and
evolution of the thermodynamic self-organised (self-
assembled), self-reproduced polyhierarchic systems."*
>   **Author**

---

[1] «Der Welt» - the Universe of R.Clausius and J.W.Gibbs - is a model of simple thermodynamic system, where no work or only extension work is performed.

CHAPTER 1

THE GENERAL SCOPE OF THE EVOLUTION OF LIVING SYSTEMS

1. THERMODYNAMIC METHODS

In this book - «Thermodynamic Theory of the Evolution of Living Beings» - we consider the theory of the evolution of living systems from the viewpoint of one of the physical models of the origin and development of life.

The notion «thermodynamics» used in the book relates to physics (and also physical chemistry and other interdisciplinary areas) that study the most general properties of macroscopic systems in the state of thermodynamic equilibrium and the transitions between such equilibrium states. The terms «thermodynamic» and «thermostatic» are equivalent in the framework of this book. Thermodynamics (thermostatics, or equilibrium thermodynamics) does not describe any processes as time dependencies. Instead, it provides exact relationships between the measurable properties of a system and answers the question how far the process (the evolution) will go until the equilibrium is achieved. Thermodynamic methods do not require any assumptions concerning the details of the system structure (molecules, cells, organisms, populations, etc.) or concerning the mechanisms that drive the system to equilibrium. The thermodynamics of a system, which forms the basis of the present work, considers only the initial and final states and is not interested whether the process under study occurs under equilibrium or non-equilibrium conditions. Therefore, it sets the time axis but does not use time as a parameter.

However, a researcher can be interested in the changes that take place in a system (for instance, a chemical or biological one) from the viewpoint of the thermodynamics of structures - systems but not of processes.

Let us denote the Gibbs function (the Gibbs free energy) of the system as G. Then the Gibbs function of *the system* (*structure*) *formation* can be represented as ΔG . One can assume that the Gibbs function has actual physical sense for simple and complex living systems (see Appendix 2) at any moment of the evolution, and its value is determined by the characteristic parameters of the state of the system (G.Lewis, M.Rendall, de Donder, K.Denbigh).

During the evolution of an open heterogeneous biological system, such as the biotissue of a living being, variation (Δ) of the specific Gibbs function of the structure formation, $\Delta\Delta\tilde{\bar{G}}$, is first of all connected with the changes in the chemical (*ch*) - molecular (*m*), and supramolecular (*im*) composition of the system. Note that the symbol "—" over G means that we consider the specific value of G, $\bar{G}$; using the symbol "~" we stress that the system is heterogeneous.

The specific value of $\tilde{G}$, $\bar{\bar{G}}$ can be taken for a unit volume or mass of the system. Variations of the Gibbs function of the system at any stage of the evolution (for instance, ontogenesis and philogenesis) can be calculated by means of thermodynamic methods, which do not, as we have already mentioned, consider time (t) as a parameter. It follows that while constructing the functions of the type $\overline{\widetilde{G}}_{system} = f(t)$ or $\Delta\overline{\widetilde{G}}_{system\text{-}formation} = f(t)$, one can use the data «obtained» by thermodynamics only to estimate the state of the system of the given composition at some fixed moment. (Fixing the time moment, we also fix a certain stage of the extent of process, ξ.) It should be taken into account that in practice one uses not the absolute values of $\overline{\widetilde{G}}$, which cannot be measured, but their variations, $\Delta\overline{\widetilde{G}}$. As to the variations of $\Delta\overline{\widetilde{G}}$, they are denoted as $\Delta\Delta\overline{\widetilde{G}}$. Sometimes the words «variation», «change», etc., are used instead of the symbol Δ. For these reasons, it is always necessary to understand what mentioned values are meant.

The specific Gibbs function of the structure formation ($\Delta\overline{\widetilde{G}}$) calculated with respect to some conventional standard level, is the measure of the thermodynamic stability of this structure (chemical, supramolecular, etc.). Its variation, $\Delta\Delta\overline{\widetilde{G}}$, characterises the changes in this stability. For instance, $\Delta\overline{\widetilde{G}}^{ch}$ ($\Delta\overline{\widetilde{G}}^{m}$) determines the stability of chemical compounds – molecules while $\Delta\overline{\widetilde{G}}^{im}$ the stability of supramolecular structures (for instance, in the state of an ideal gas or in condensed phase, respectively).

Variations of the functions $\overline{\widetilde{G}}^{ch}$ and $\overline{\widetilde{G}}^{im}$ (we stress once more that in practice one deals with the variations of $\Delta\overline{\widetilde{G}}^{ch}$ and $\Delta\overline{\widetilde{G}}^{im}$) can be connected for an open system with two factors (1 and 2).

Indeed, it is convenient to consider an open supramolecular structure as two subsystems functioning separately but coinciding in space. One of the subsystems (1) is treated as a closed (thermodynamically quasi-closed) one. Variations of $\Delta\overline{\widetilde{G}}_1^{im}$ ($\Delta\Delta\overline{\widetilde{G}}_1^{im}$) in this system are caused by rapid processes, which can be assumed to take place in a system with constant composition. The other subsystem, 2, is an open one, and the variation of $\Delta\overline{\widetilde{G}}_2^{im}$ ($\Delta\Delta\overline{\widetilde{G}}_2^{im}$) occurs due to the changes in the composition of its chemical substance. This slow process is related to the changes in the nature and the amount of supramolecular structures of the open system, which exchanges chemical substances with the environment. This second subsystem can be considered as partly closed kinetically (quasi-closed kinetically) since it accumulates supramolecular structures with higher stability.

There also exist other models that decompose the processes taking place in open systems in separate components.

Thus, thermodynamics is able to provide the data for constructing functions of the type $\Delta G_{system\ formation}=f(t)$, which should be studied by means of kinetic methods. It is worth stressing once more that the theoretical model of the evolution presented in the work describes only variation of the state functions relating to the *systems*. Naturally, it cannot be used to study molecular or any other mechanisms of the processes that change the state of the systems. As to the *processes*, which are mostly non-equilibrium, they are investigated both by the kinetics and the thermodynamics of non-equilibrium processes (more precisely, by the «kinetic thermodynamics»). Note that one usually distinguishes between the two cases. First, there exists thermodynamics of non-equilibrium processes that take place under the conditions close to equilibrium. The study of such processes can be based on the general rules of thermodynamics - thermostatics. The other case is the thermodynamics (more accurately, kinetics) of non-equilibrium processes that run under the conditions far from equilibrium. In this case, strictly speaking, the systems cannot be characterised by state functions, i.e., by functions whose differentials are full ones. One can also distinguish the thermodynamics of equilibrium processes as a separate field of studies. However, in the present work we are mostly interested in the thermodynamics (thermostatics) of systems.

2. EVOLUTION

The term «evolution» is used in the present book, as a rule, in its common sense. The biological evolution is considered as an irreversible process of the historical variation of life with respect to the evolution time scale characterising the given object. However, in contrast to the evolution theory by Ch.Darwin (Ch.Darwin and A.Wallace) or to the modern synthetic theory of evolution, which both emphasise the evolution of populations and are in fact kinetic theories, the physical phenomenological theory presented in this book points out only the thermodynamic tendency of the evolution and the extent of its progress.

The theory by Ch.Darwin is commonly referred to as a general theory that studies the causes, «kinetic» motive forces, mechanisms, and the general rules of the evolution of living organisms from biological point of view. This theory forms the theoretical basis of all biology.

The development of modern darwinism is connected with the analysis of the data obtained in molecular biology. It is aimed at a more profound understanding of hereditary variability and at finding the ways to control living natural resources.

However, up to now the theory of biological evolution did not attempt to point out the physical essence of the evolution tendency. Though, Ch.Darwin admitted that «the principles of life are a part or a consequence of some general law» determining the evolution of matter as a whole. The author supposes that life is a particular manifestation of the general natural laws.

The thermodynamic theory of the evolution of living beings on the Earth considers the smooth transition from the chemical evolution to the biological one (the 'continuity principles', according to Ch.Darwin and M.Kalwin) based on the hierarchic model of living and non-living substance organisation. In spite of the fact that this model, based on hierarchic thermodynamics, can be in principle applied to the evolution of populations and human society, the author restricts himself to the discussion of chemical and supramolecular aspects of the evolution. Chemical substance is the «universal» material of the biological evolution used by nature on all hierarchic structure levels.

Probably, in future the thermodynamic theory of biological evolution will «merge» with Darwin's theory.

In any case, there is no contradiction between these two theories, and one can hope that in future the thermodynamic theory will help to «pacify» a lot of supporters and opponents of Darwin's theory on the grounds of natural science.

3. THE ORIGIN OF LIFE AND THE ROLE OF DIVINE INITIATION

Most of the investigators suppose that life appeared on the Earth in a natural way and the biological evolution is the continuation of the chemical substance evolution. They try to explain the origin and development of life on our planet in the framework of the general natural laws.

The general natural laws conventionally include the energy conservation law, the first and the second laws of thermodynamics. (The first law of thermodynamics is a particular case of the energy conservation law.) The statistical laws are closely related to this group. Nowadays, an educated person has a steady confidence in all these laws.

The general laws cannot be derived from any «elementary» statements. They are simply given to us by Nature. It is at this stage of studying our world that we come to the concept of God. However, this concept does not mean the same for different people. On the one hand, many professional scientists seem to understand «God» as a symbol of Nature together with its constant general laws, which are given to human beings. On the other hand, for most of the people God, created by man in his own image, is the Creator of everything surrounding us and happening in the Universe.

Probably, the absolute majority of the people turn towards God, many of them create for the sake and glory of God. Nevertheless, each one has «his own God»!

The author tries to understand the evolution of the matter (first of all, chemical and biological evolution) using as the starting point only the general natural laws, without the concept of God as a living being accessible to our imagination. The «lucky star» of the author while developing the model of biological evolution was his strong belief in the general laws and exact theories created by R.Clausius, J.W.Gibbs, and other classics of modern natural science. The suggested model is based on distinguishing between the thermodynamics and the kinetics, and it does not utilise the concept of the dynamic self-organisation (or simply self-organisation, in terms of I.Prigogine), which has widely spread during the second part of this century.

4. Several Facts that are to be Explained by the Physical Theory of Biological Evolution

There are plenty of commonly known facts in the biological world whose explanation is not subject to the theory by Ch.Darwin. Nevertheless, some phenomena can be studied from the evolution viewpoint in the framework of the general natural principles. Let us consider several known dependencies and facts.

1. It is known that the chemical composition of living objects, such as, for instance, animals and plants, varies in the course of ontogenesis, philogenesis and at long-lasting stages of the

biological evolution. During the evolution development the biosystems (organells, cells, the biotissue of an organism, the biomass of a population, etc.) get enriched with energy-intensive substances, which oust water from these systems. Such substances are mostly organic compounds - lipids (fats), proteins, polysugars, nucleic acids, and so on. For instance, immediately after conception a human embryo contains at least 95-97% of water, while the tissues of an elderly person are poor in water - its content is reduced to 60-65%.

Variation of the composition of tissues during the ontogenesis is especially noticeable for young plants and animals. There are considerable composition variations during the process of embryo development, for instance, for mammals. Fig.1 shows the results of studying the variation of water and fat content in the developing human embryo. Fig 2 presents the variation of the amount of water and fat in the tissues of ageing rats. The data presented in Fig.3 show the variation of water and protein content in the tissues of cattle. The data was obtained both for the developing embryos and for young animals after birth. Similar examples can be easily found in original papers and reference literature.

As to the philogenesis and the changes in the chemical composition accompanying it, the situation resembles the one with the ontogenesis. Science has revealed multiple examples of similar behaviour.

Analysing the experimental results one comes to the conclusion that both in the ontogenesis and in the philogenesis the energy capacity of a system grows and its chemical thermal stability decreases. This, as the author has shown, is accompanied by the increase of the thermodynamic stability of the supramolecular structures of the tissues of organisms. This rule is schematically illustrated in Fig.4 for the example of ontogenesis.

Clearly, a theory is needed to explain the above-listed facts!

2. It is known that the chemical composition of living beings adapts to the conditions of their habitat, such as food, temperature (T), pressure (p), and other parameters (V.Aleksandrov, J.Somero, and others). Probably, the most bright examples are the changes in the composition of fatty acids (fats) in the cells of microorganisms, plants, and animals due to the temperature variation of their environment. When the temperature of an organism is lower than the optimal one, then there is an increase of the amount of unsaturated fatty acids; in the opposite case - at high temperatures - the fats of living organisms are enriched with saturated fatty-acid radicals.

The data in Fig.5 relates to the iodine number, N, which is an unambiguous characteristics of the fatty acids unsaturation level: larger N correspond to higher unsaturation levels. The plot in Fig.5 demonstrates how the iodine number of the phosphatides (lipids) of the fly larvae depend on the temperature of their maturation (habitat). Can this fact, as well as numerous analogous ones, be explained in the framework of thermodynamics? The answer turns out to be positive.

3. There is a lot of experimental evidence relating to the variation of melting (denaturation) temperature of animal chromatin, as well as fat, collagen, and other tissues, in the course of ontogenesis and philogenesis of organisms. All these facts can be also explained from the viewpoint of thermodynamics.

4. The well-known empirical law by Weber-Fechner describes the reaction (reply) of living objects to various external influences. Indeed, light, acoustic and magnetic fields, mechanical torsions, smells, and psychological stresses initiate the reaction of biological objects and determine their behaviour.

Many facts in the behaviour of living beings have been already explained with the help of the Le Chatelier - Brown principle. However, this principle can be applied only to closed systems. Thermodynamical theory of behaviour, which studies quasi-closed systems, «derives» the Weber-Fechner law and can provide quantitative estimations whether the Le Chatelier - Brown principle can be applied to actual biosystems or not.

5. The optical activity of biological substances is one of the principal peculiarities of living world. Therefore, those researchers who consider the origin of life as a natural process are eager to explain the formation of optical-active compounds by means of physical theories. One of such approaches is to suppose that the optical activity, at least on the Earth and in the Solar System, has thermodynamical origin. For instance, according to some works published by the author, the principal role here is played by natural physical fields and the Coriolis forces, whose influence manifests itself only on the level of macroscopic objects. We shall not discuss here this viewpoint since so far there are too little experimental facts confirming it. The problem is still waiting for its solution.

6. Is it possible for thermodynamic properties to be inherited? The factual data, at least on the molecular

level, allow the answer to be positive. However, more detailed studies are necessary in this area.

5. POINTING PUT NEW CONFORMITIES

The biological evolution can be understood in the framework of the general laws of nature if one takes into account the hierarchic structure of biological objects and finds the way to choose quasi-closed (both thermodynamically and kinetically) thermodynamic systems. The evolution model can be developed on the basis of hierarchic thermodynamics, or macrothermodynamics. The hierarchic thermodynamics manifests itself in the natural law (most probably, a general one) revealed by the author: there exist unidirected seria of life-spans (times of relaxation or existence) for structures of different hierarchies (Fig.6).

For instance, if we consider the biotissue cells of some organism, the organism itself, and the population formed by such organisms (i.e., a fragment of the hierarchic sequence of biostructures), the following relation can be observed: the average life-span (t) of a cell (*cel*) in the organism is much less than the average life-span of the organism (*org*), which, in its turn, is much less than the life-span of the population (*pop*):

$$\ldots << t^{cel} << t^{org} << t^{pop} <<\ldots \tag{1}$$

The law (1) can be laconically formulated as follows: *structures of lower hierarchy (j) live (exist) in biosystems much shorter than structures of higher hierarchy (j+1).*

The series (1), in a sense, «sets the rules» for the formation of self-reproducing identical polyhierarchic structures. Each higher structural hierarchy creates the habitat (the thermostat, in a broad sense, - see Appendix 2) for all lower-hierarchy structures. If time hierarchies did not exist, the substance in such a world would stay in the state of «homogeneous-heterogeneous mixture», and there simply would be no phenomenon of life.

This law was applied for the study of the molecular and supramolecular hierarchies (molecular systems and systems of supramolecular structures, respectively) in various biological objects. Investigations showed that one can partly «close» the thermodynamic systems under study (the biological objects or their parts) and describe their evolution in terms of state functions whose specific values tend to their extreme values.

It should be noted that the fundamental role of time hierarchies in physics has been pointed out by N.Bogolubov, and the principal possibility to build temporal organisation in biology has been demonstrated by H.Kacser, S.Waddington, and especially B.Goodwin. Later the author of the present work found out that there exist long uni-directed time sequences corresponding to the seria of hierarchic structures, which are ordered according to their sizes or formation energies.

The existence of the series (1) dividing the time axis into components means, in fact, that the structures of lower hierarchies (say, the cells) undergo evolutionary changes at times much less than the life-span of the organism. The biomass of the organism is a kind of environment, whose parameters are almost constant at small times. The evolution of cells is a non-stationary process; it takes place against the background of chemical compounds coming into the cell, so that the flow has practically constant composition. In other words, the cell subsystem, as we have already mentioned, has its own thermostat (see Appendix 2), and this enables us to consider this subsystem as a quasi-closed one with respect to certain parameters.

Certainly, each biosystem (organell, cell, organ, organism, species and so on) has its own thermostat (habitat) with its own essential parameters. A system, together with its thermostat forms a complete thermodynamic system. Combined together, such complete systems form a new system of higher hierarchic level, which, in their turn, have its own thermostat, and so on.

A given system j of any hierarchy, as well as any thermostat, which is also considered as a system, $j+1$ or $j+n$, where $n = 2,3....$ has a tendency to the increase of its superstructure thermodynamical stability. The superstructure is formed as a result of the interaction between the elements $(j-1)$- unitary structures of the hierarchy (j) under study. When molecular (chemical) systems are investigated, one considers formation of supramolecular structures due to the aggregation of molecules, supramolecular aggregates, cells, ets.

There always exist fluctuations of the thermostat parameters at corresponding relatively short time intervals. For instance, there can occur a change in the food composition of an organism in the course of ontogenesis or a considerable change of the environmental temperature. As a result, for the j-th thermostat of any hierarchy, the value of $\bar{\bar{G}}_j$ is varying and can be called as "floating value". However, in the life time scale of the system (j-1) these variations are averaged ($\overline{\overline{G}}_j \approx$ constant) and minimum of $\overline{\widetilde{G}}_{j-1}$ is observed (Fig.4).

Thus, it turns out that there exists a thermodynamic model of the evolution of living objects that allows to reveal the changes taking place in these objects - the same way as chemical thermodynamics deals with physical-chemical systems where processes analogous to heterophase polymerisation and so on take place (see Chapter II).

It should be stressed that the model we discuss is exact only by definition. Therefore, the next step should be aimed at finding out whether the model fits the reality. The author shows that within the framework of the assumptions we use (connected, first of all, with the strong inequalities for the series similar to (1)) the model agrees with the facts observed. Several calculations carried out by the author and presented here as examples will allow the reader to see some advantages of the model.

What possibilities for the study of the world does the suggested theory provide? It is difficult to answer this question in detail. However, there are grounds to suppose that these possibilities will be restricted only by the thermodynamic method itself.

While developing the model, the author relied on the general principles of science, according to which the models can be quite different, depending on the aims of investigation, the chosen time scale, and the volume of the object under study. According to the facts presented above, it seems difficult to agree with the extreme opinion of some researchers who neglect the study of the evolution based on the general natural laws. Instead, they make categorical claims, such as that the evolution of natural systems «is governed by the flows of energy and substance passing through them». To be sure, this statement is reasonable only from the kinetic point of view. As a rule, it relates only to the models of a single type, based on the concept of dynamic self-organisation.

In this book we do not discuss all models of evolution known at present. The author only tries to draw attention (of the young scientists, first of all) to the following fact: there exists a physical theory that provides a satisfactory explanation for the origin of living beings and their evolution (without considering its mechanisms). The theory is not connected with the fashionable concept of dynamic self-organisation, the ideas of «self-organised critical point» and many others. Time will show which concepts are more promising. The role of the judge will be played by strict quantitative experimental test, which will check the validity and universality of the theory.

Science became a public phenomenon. Nevertheless, there seem to exist some researchers who disregard the commonly accepted view on science as an object for human activity.

The next section, to a certain extent, reflects the author's opinion concerning this question.

6. THE ASSUMPTIONS, THE MODEL AND THE REALITY

Strictly speaking, all natural processes are more or less irreversible. However, it is well known that science widely uses the ideas of classical thermodynamics, which deals only with equilibrium states and reversible (equilibrium) processes. The point is that very often real systems can be considered as equilibrium and some processes and transformations can be, as a certain approximation, studied as reversible ones. A researcher would also like to know to what approximation natural processes and systems can be considered as ideal and studied in the framework of «ideal» classical thermodynamics. While dealing with such problems, one sometimes speaks about «the limits of gnoseological accuracy». Within these limits, thermodynamic effects can be defined as equilibrium or non-equilibrium, stable or nonstable (L.Sedov).

The equilibrium state of a system can be stable or nonstable. Sometimes, the state of nonstable thermodynamic equilibrium, or metastable equilibrium (quasi-equilibrium) can be treated as an equilibrium state. This is possible when the process under study, which is connected with the variation of some parameter, is stable and its typical (characteristic) times are much longer or much shorter than the times corresponding to other processes in the system. It is also necessary that there are no noticeable changes in the environment at times typical for the chosen process. In other words, one and the same process can be considered as equilibrium or non-equilibrium, depending on the chosen time scale. According to the problem that is to be solved, systems and processes can be described in different ways.

Thus, in any particular case a given state of the system can be considered as equilibrium or non-equilibrium to different orders of approximation.

The last statement becomes clear if we take into account that while studying a natural phenomenon one always uses the model that fits the reality only to a certain extent. There is no contradiction here; it is worth recalling the experience of physics. Indeed, one and the same phenomenon can be treated by different theories as an equilibrium or non-equilibrium one. For instance, it is well known that quasi-equilibrium processes, such as chemical transformations, plastic deformations, electromagnetic hysteresis, etc., are regarded as irreversible. However, in corresponding time scales this irreversibility is often neglected. It should be noted once more that there is no ideal equilibrium and complete reversibility in Nature.

The same situation is observed when the dynamics (kinetics) of processes is studied with respect to stationarity and non-stationarity. The term «stationary state» usually means that some essential parameters determining the behaviour of the system do not vary in time. If these parameters vary with time slowly, the state is called quasi-stationary. When the resulting flow coming into the system tends to zero, the system achieves equilibrium. The method of stationary state often allows to simplify the study of the systems and for this reason it is widely used in natural sciences. For instance, theorems by L.Onsager and I.Prigogine imply the existence of a stationary state for the systems. These theorems are valid for some open systems close to equilibrium, but the evolution of real biological non-stationary systems cannot be understood with their help. Thus, even the kinetic description of real systems implies certain restrictions on approximating phenomena by means of stationary models.

7. THE MODEL OF BIOLOGICAL EVOLUTION

Due to the action of the solar energy, the substances that were thermodynamically stable initially transform (as nowadays) into various products of photosynthesis (Lord Porter). Then, as a result of spontaneous «dark reactions» these products transform into various substances according to the laws of chemical thermodynamics. The choice of these substances is «made» by kinetics in correspondence with the laws of chemical thermodynamics. According to the laws of the local thermodynamics of supramolecular structures, the most stable superstructures are selected from the whole spectrum of chemical substances, owing to the fact that $\overline{\overline{G}}{}^{im}$ for biostructures tends to a minimum. These stable superstructures are accumulated in the micro- and macro-objects of the system (G.Gladyshev). Some macromolecules and superstructures get reduplicated, due to possible matrix mechanisms.

The first to be selected are nucleic acids, whose composition and structure (because of the thermodynamic factors) slowly adapt to the nature of the surroundings including the nature of proteins which in turn is determined by the DNA structure. This explains the existence of feedback between the structure of protein and the DNA. According to our model, this feedback has thermodynamic origin.

The processes of DNA proceed in parallel with the degradation processes - the decay of chemical substances. However, living systems oppose this and tend to preserve their state. This tendency is thermodynamic in nature. The biosystems reproduce the decomposing supramolecular structures, and, as we have already noted, thermodynamics chooses the most stable ones.

It is thermodynamically beneficial for macromolecular chains to pair with similar chains and to surround themselves, by means of supramolecular contacts, with the renewed «young» substance of living organisms. Therefore, the evolution selects those thermodynamically preferable paths which promote the division of cells and preservation of the DNA. All this takes place against the background of the fluctuations in the parameters of the thermostat (the surroundings). Together with other factors, this ensures the retaining of life. Thermodynamical factors help to stabilise all complex biological structures and lead to the appearance of higher hierarchies in the biological world.

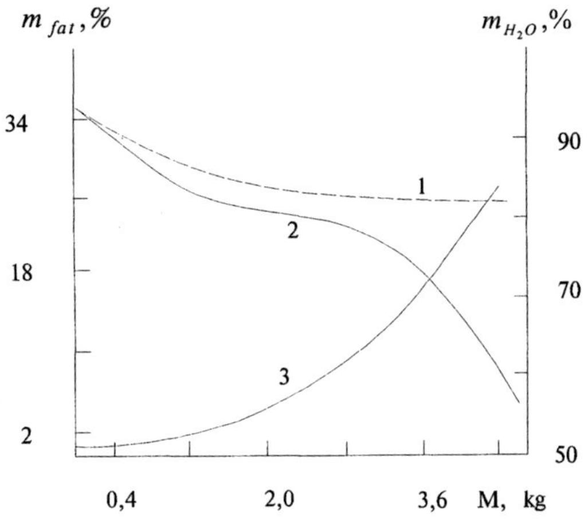

Fig.1. Variation of the amount of water and fat in a developing human embryo (*Widdowson E.M. Body Composition in Animals and Man*, 1967).

1 - water in fat-free tissue; 2 - water in tissue; 3 - fat in tissue. m_{fat} and m_{H_2O} - the amount of fat and water (weight %); M - mass of embryo.

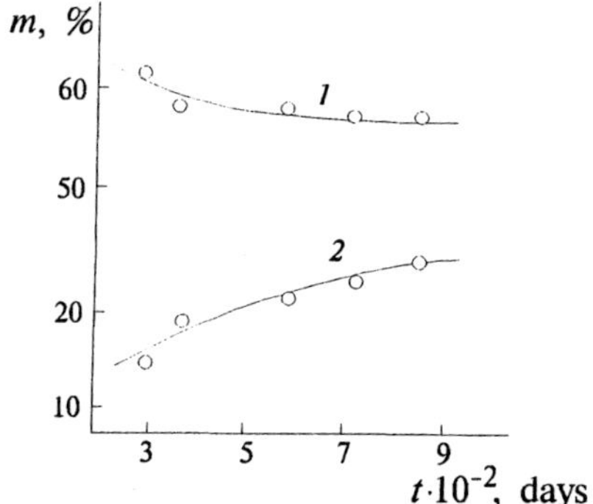

Fig.2. Variation of the amount of water (1) and fat (2) in the tissues of aging rats (Lesser G.T., et. al., 1980).

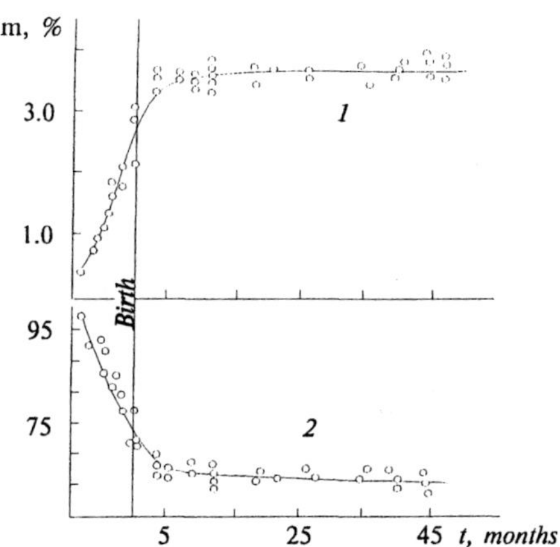

Fig.3. Variation in the contents of nitrogen (protein) (1) and water (2) during ontogenesis for cattle and human (*Moulton C.R.,* 1923).

The experimental data for the embryos relate to the cattle and human. The data on the postbursal period were obtained for tissues of the cattle.

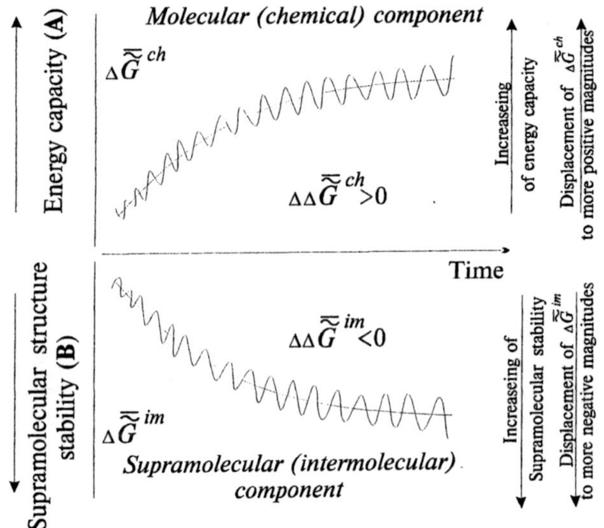

Fig.4. Schematic plot depicting the variation of the specific chemical energy capacity of the biomass - A ($\Delta\overline{\overline{G}}^{ch}$, or $\Delta\overline{\overline{H}}^{ch}_{comb}$) and the thermodynamic stability of its supramolecular structures during the ontogenesis of living beings - B ($\Delta\overline{\overline{G}}^{im}$). The arrows near the axes (A) and (B) point the direction in which the energetic capacity and supramolecular stability increase. The growth of $\Delta\overline{\overline{G}}^{ch}$ ($\Delta\Delta\overline{\overline{G}}^{ch} > 0$) means that the energetic capacity of the system increases; displacement of $\Delta\overline{\overline{G}}^{im}$ to more negative values ($\Delta\Delta\overline{\overline{G}}^{im} < 0$) means that there is an increase of the supramolecular stability of the system.

The value A corresponds to the variation of $\overline{\overline{G}}^{ch}(\Delta\overline{\overline{G}}^{ch} \equiv \Delta\overline{\overline{G}}^{m})$ due to the chemical substance formation from chemical elements or simple substances (the chemical or molecular component). The value B corresponds to the variation of $\overline{\overline{G}}^{im}$ ($\Delta\overline{\overline{G}}^{im}$) due to the supramolecular structure formation in the course of self-assembly (the supramolecular component). The plots for A and B have different scales. The time axis is set by the second law of thermodynamics and has no scale. $\Delta\overline{\overline{G}}^{0}_{f} = \Delta\overline{\overline{G}}^{ch^0}_{f} + \Delta\overline{\overline{G}}^{im^0}_{f}$ (see Appendix 2), $\Delta\overline{\overline{G}}^{0}_{f}$ - standard Gibbs function of the structure formation; $\Delta\overline{\overline{G}}^{ch^0}_{f}$ is much larger than $\Delta\overline{\overline{G}}^{im^0}_{f}$.

The specific values of $\Delta\overline{\overline{G}}^{ch}$ and $\Delta\overline{\overline{G}}^{im}$ can be defined for unit volume or mass of the system. In both cases the dependencies shown in the figure are similar.

Usually in chemical thermodynamics the abscissas are given not by the values of time, t but the extent of the process accomplishment, ξ.

The saw-tooth lines plotted against the curves emphasize that the fluctuation of the parameters of the surrounding, such as temperature, pressure, nutrition schedule, physical fields, the change of day and night, the change of seasons, etc., lead to the variations of $\Delta\overline{\overline{G}}^{ch}$ and $\Delta\overline{\overline{G}}^{im}$. The organism adapts to these variations only within the limits of the adaptive zone.

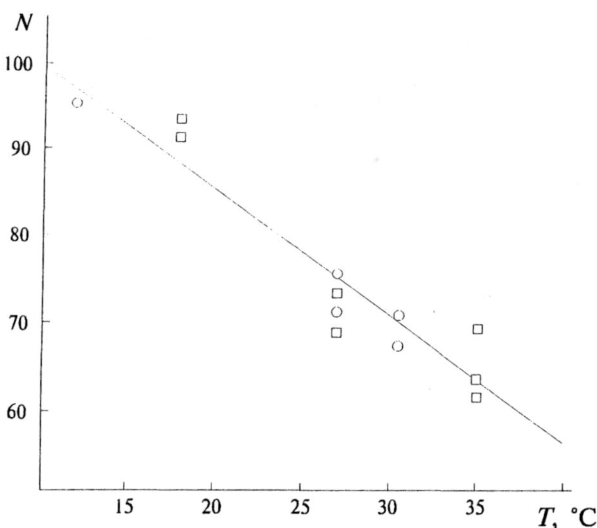

Fig.5. The iodine number of phosphatides for larvae *Calliphora erythrocephala* (□) and *Phormia terranova* (o) versus the environmental temperature during the maturation of the flies (*Fraenkel, Hoppe,* 1940; *Aleksandrov,* 1975).

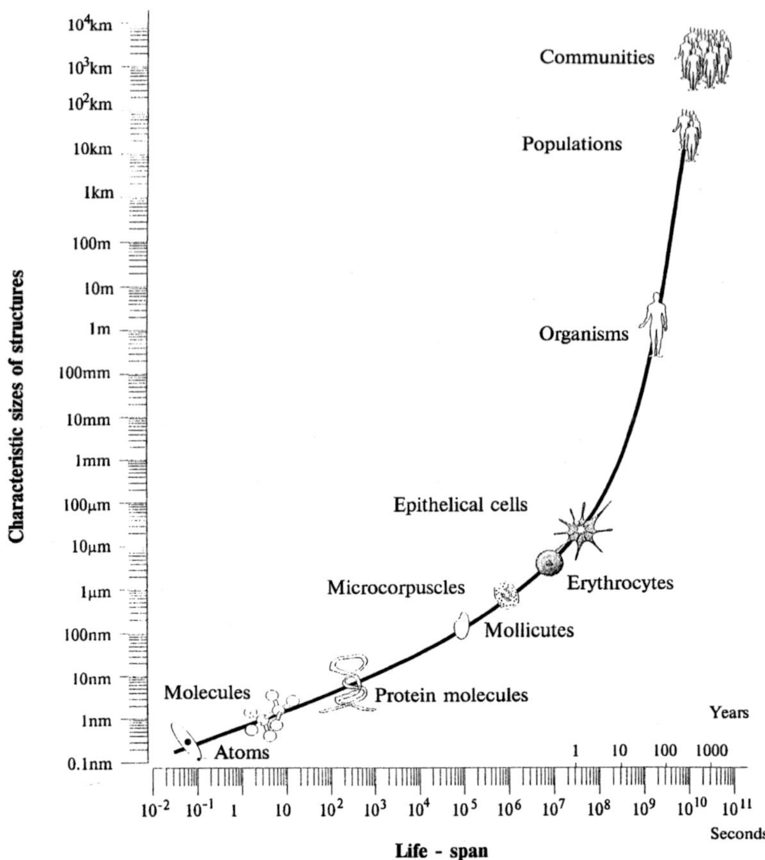

Fig.6. Relation between the characteristic sizes (*l*) of structures belonging to different hierarchies and their life-spans (*t*) in the biomass.

The dependence is presented in the logarithmic scale. The scheme is based on the facts relating to Homo sapiens.

REFERENCES

1.*Alberty R.A.* (1987). *Physical Chemistry.* 7th Ed. New York, etc.: Wiley, 934 p.

2.*Alberts B., Bray D., Lewis J., Raff M., Roberts K., Watson J.D.* (1989). *Molecular Biology of the Cell.* 2nd Ed. New York - London: Garland Publ. Inc.

3.*Aleksandrov V.Ja.* (1975). *Kletki, Makromolekuly i Temperatura* (Cells, Macro-molecules, and Temperature). Leningrad: Nauka, 330 p.

4.*Bazarov I.P.* (1983). *Termodinamika.* (Thermodynamics). 3rd Ed. Moscow: Vysshaya shkola, 344 p.

5.*Body Composition in Animals and Man.* (1968). Proc. Symp. Held May 4-6, 1967. Univ. of Missouri. Columbia Publ. 1598. Wash. (D.C.): Nat. Acad. Sci., 521 p.

6.*Bonner W.A.* (1991). *Biomolecular Chirality. Origin of Life and Evol.* Biosph., v. 21, p.59.

7.*Broda E.* (1975). *The Evolution of the Bioenergetic Processes,* Oxford, New York, Toronto, Sydney, Paris, Braunschweig: Pergamon Press, 300 p.

8.*Calvin M.* (1969). *Chemical Evolution.* Oxford: At Clarendon Press.

9.*Cantor C.R., Schimmel P.R.* (1980). *Biophysical Chemistry.* 3 v. San Francisco: W.H. Freeman and Co.

10.*Chela-Flores J.* Ed. (1996). *Physics of the Origin and Evolution of Life.* Dordrecht, Boston, London: Kluwer Acad. Publ. (in press).

11.*Clausius R.* (1865). *Pogg. Ann. Bd.* 125, s.400.

12.*Darwin Ch., Wallace A.R.* (1858). *On the Tendency of the Species to Form Varieties and on the*

Perpetuation of the Species by Natural Means of Selection. J.Linn.Soc. (Zoology), v.3, p.45.

13. *Davson N., Danielli J.F.* (1970). *The Permeability of Natural Membranes.* 2nd Ed. Darien: Hafner Publ. Co. 365 p.

14. *Deamer D.W., Fleischaker G.R.* (Ed.) (1994). *Origins of Life. The Central Concepts.* Boston - London: Jones and Bartlett Publ. Inc.

15. *Denbigh K.G.* (1975). *An Inventive Universe.* London: Hutchinson, 220 p.

16. *Denbigh K.G.* (1989). Note on Entropy, Disorder and Disorganization. *Brit. Jour. Phil. Sci.,* v.40, p.323.

17. *Denbigh K.G.* (1989). The Many Faces of Irreversibility. *Brit. Jour. Phil. Sci.,* v.40, p.501.

18. *Denbigh K.G.* (1971). *The Principles of Chemical Equilibrium.* 3rd Ed Cambridge; Cambridge Univ. Press, 491 p.

19. *Denbigh K.G.* (1951). *Thermodynamics of the Steady State.* London: Methuen, 103 p.

20. *Edsall J.T., Gutfreund H.* (1983). *Biothermodynamics.* Chichester, New York, Brisbane, Toronto, Singapore: John Wiley & Sons.

21. *Flint Von Rainer* (1988). *Biologie in Zahlen.* Stuttgart, New York: Gustav Fischer Verlag.

22. *Fox S.W. and Dose K.* (1977). *Molecular Evolution and the Origin of Life.* Revised Ed. New York: Marcel Dekker.

23. *Gerasimov Ya.* (Ed.) (1985). *Physical Chemistry.* 1,2 v. M.:Mir Publ., 1230 p.

24. *Gibbs J.W.* (1928). *The Collected Works of J. Willard Gibbs. Thermodynamics.* V.1. New York: Longmans, Green and Co., 55-349 p.

25.*Gladyshev G.P.* (1994). The Motive Force of
Biological Evolution. Vestnik RAN, v.64, N 3,
p.221. (Herald of the Russian of Sci., v. 64. N 2,
p. 118).

26.*Gladyshev G.P.* (1995). Termodinamika of the
Ierarkhicheskih Sistem (Thermodynamics of the
Hierarchic Systems). In: *Khimicheskaya Ensiklopedia*
(Chemical Encyclopedia). M.: v.4, p.535.

27.*Gladyshev G.P.* (1978). On the Mechanisms of
Reactions in Rarefied Gas: Processes in the Solar
System. *Moon and Planets,* N 19, p.89.

28.*Gladyshev G.P.* (1978). On the Thermodynamics of
Biological Evolution. *J.Theoret.Biol.,* v.75, p.425
(Preprint, Chernogolovka, Inst. Chem. Phys., 1977,
p.46)

29.*Gladyshev G.P., Khasanov M.M.* (1981). Optical
Activity and Evolution. J. Theor. Biol., v.90,
p.191.

30.*Gladyshev G.P.* (1992). Oriented Hydrodynamic Flows
in Rotational Medium and Assymetry in Bioworld.
Ukrainian Polymer J., v.1, N 1, p.55.

31.*Gladyshev G.P.* (1995). Thermodynamic Trends of
Biological Evolution. *Izvestia RAN. Seria biol.,*
1995, N 1, p.5 . (Biology Bulletin, v.22, N 1, p.1.
ISSN 1062-3590; N 1, 1995).

32.*Gladyshev G.P.* (1996). Thermodynamic Trends of
Biological Evolution. Model and Reality. *Biology
Bulletin* ISSN 1062-3590, N4, 1996.

33.*Gladyshev G.P.* (1988). *Termodinamika i Makrokinetika
Prirodnykh Ierarkhicheskih Processov* (Thermodynamics
and Macrokinetics of Natural Hierarchic Processes).
Moscow: Nauka, 287 p.

34.*Gladyshev G.P., et. al.* (1996). Thermodynamics of
Biological Evolution. *J. Biol. Phys.* (in press).

35.*Gladyshev G.P.* (1996). Thermodynamics and the Evolution of Biological Structures. *Biofizika* (in press).

36.*Gladyshev G.P. and Komarov F.I.* (1996). Hierarchic Thermodynamics and Gerontology. *Vestnik Ross. Med. Acad.*, N 6, p. 31.

37.*Guggenheim E.A.* (1977). *Thermodynamics: An Advanced Treatment for Chemists and Physicists.* 6th ed. Amsterdam etc.: North Holland, 390 p.

38.*Goodwin B.C.* (1963). *Temporal Organisation in Cells.* London and New York: Academic Press.

39.*Green N.P.O., Stout G.W., Taylor D.J., Soper R.* (Ed.) (1989). *Biological Science.* Cambridge: Cambridge Univ.Press. 9th printing.

40.*Haywood R.W.* (1980). *Equilibrium Thermodynamics.* Chichester, New York, Brisbane, Toronto: A Wiley Interscience Publ.

41.*Jones M.N.* (Ed.) (1979). *Biochemical Thermodynamics.* Amsterdam, Oxford, New York: Elsevier Sci. Publ. Co.

42.*Hochachka P.W., Somero G.N.* (1973). *Strategies of Biochemical Adaptation.* Philadelphia, London, Toronto: W.B.Saunders Co.

43.*Kacser H.* (1957). *The Strategy of the Genes.* London: Allen and Unwin, p. 191.

44.*Kubo R.* (1968). *Thermodynamics.* Amsterdam: North-Holland Publ. Co.

45.*Kanungo M.* (1982). *Biokhimiya Stareniya.* Moscow: Mir, 294 p.

46.*Lauffar M.A.* (1975). *Entropy-driven Processes in Biology.* Boston: Springer, 261 p.

47.*Lesser G.T., Deutch S., Markofsky J.* (1980). *Amer. J. Physiol.*, v.238, N 7, Pt.1, p.R82-R90.

48.*Lewis G.N., Rendall M.* (1961). *Thermodynamics.* 2nd Ed. N.Y.: McCraw-Hill, 723 p.

49. *Mitchell P.* (1979). Keilin's Respiratory Chain Concept and Its Chemiosmotic Consequences. *Science,* v.206, p. 1148.

50. *Morgan T.H.* (1935). *The Scientific Basis of Evolution.* New York: Norton, 300 p.

51. *Moulton C.R.* (1923). *J. Biol. Chem,* v. 57, p.79.

52. *Novak V.J.A.* (1992). The Role of Neoteny and Sociogenesisin the Evolution of Cell Structure. *Trieste Conference on Chemical Composition and the Origin* of Life, 26-30 October 1992. Trieste (Italy). Hampton, Virginia: A. DEEPAK Publ. 1993, p.321.

53. *Oparin A.I.* (1957). *The Origin of Life on Earth.* New York: Acad. Press. Inc., 495 p.

54. *Ponnamperuma C., Chela-Flores J.* (Ed.) (1995). *Chemical Evolution: Structure and Model of the First Cell.* Dordrecht, Boston, London: Kluwer Acad. Publ., 383 p.

55. *Porter G.* (1983). Transfer and Storage of Chemical Radiation Potential. *J. Chem. Society,* Faraday Trans, 2, <u>79</u>, p. 473.

56. *Prigogine I.* (1980). *From Being to Becoming: Time and Complexity and the Physical Sciences.* San Francisco: W.H.Freeman & Co.

57. *Ratner V.A., Zharkikh A.A., Kolchanov N.A., et.al.* (1985). *Problemy teorii molekulyarnoi evolutsii.* (Problems of Molecular Evolution Theory). Novosibirsk: Nauka, 264 p.

58. *Ricklefs R.E., Finch C.E.* (1995). *Aging. A Natural History.* N.Y.: Scientific American Library, 210 p.

59. *Rutten M.G.* (1973). *The Origin of Life.* Amsterdam-London, N.Y.: Elsevier Publ. Co

60. *Schneider Thomas D.* (1991). Theory of Molecular Machines. *J. Theor. Biol.* 148, 83-123; 148, 125-137, 1991.

61. *Sedov L.I.* (1984). Scientific Theories, Models and Reality. *Priroda.* N 11, p.3.

62. Somero G.N. (1986). Protein Adaptation and Biogeography: Threshold Effects on Molecular Evolution. *Tree, v.* 1, N 5, p. 124.

63. *Sychev V.V.* (1981). *Differentsial'nye uravneniya termodinamiki* (Diffeerntial equations of thermodynamics). M.: Nauka, 195 p.

64. *Sychev V.V.* (1986). *Slozhnye termodinamicheskie sistemy* (Complex thermodynamic systems). M.: Energoatomizdat, 208 p. (Translated into English: M.: Mir, 1985).

65. *Tamar H.* (1972). *Principles of Sensory Physiology.* Springfield: Charles and Thomas. (Russian transl.: Moscow: Mir, 1976, 520 p.).

66. *Waddington C.H.* (1957). *The Strategy of the Genes.* London: Allen and Unwin, 262p.

67. *Westerhoff Hans V., K. Van Dam* (1987). *Thermodynamics and Control of Biological Free-Energy Transduction.* Amsterdam. Elsevier. (Russian transl.: Moscow: Mir, 1992, 685 p.).

68. *Wrigt R.H.* (1964). *The Science of Smell.* London: Acad. Press. 143 p.

69. *Zommerfeld A.* (1952). *Thermodynamik und Statistik.* Wiesbaden. 480 s.

CHAPTER II

THERMODYNAMICS AND THE EVOLUTION OF BIOLOGICAL SYSTEMS

1. GENERALITIES

Even the latest achievements of modern natural science gave no commonly accepted answer to the question what is the motive force of biological evolution. Probably, one can hardly hope that this question can be solved by means of the inductive approach alone. This approach, widely used in biology, consists in accumulating separate facts in order to generalise them into some basic principles. At present, most likely, deductive approaches are more preferable for solving this problem than the inductive ones.

The thermodynamic (thermostatic) method [1] is probably one of such deductive methods. Here it should be noted that speaking about the thermodynamic approach I only mean that it can be used as a "starting-point". Nothing more! In order to achieve comparatively full understanding of the phenomenon of life (in the broad sense) one should also use dynamical methods, for they provide the information on the mechanisms of processes. In other words, both thermodynamics and kinetics are needed for the study of the world. However, taking into account the scientific achievements of our century, it is reasonable to suppose that these methods should be often used "in turn", so that thermodynamics should be "separated" from kinetics and vice versa. Besides, it is worth noting that one should strictly distinguish between the concepts of a thermodynamical "system" and of a "process" that can take place in that system. The behaviour (the evolution) of systems, as a rule, is studied by thermodynamics, while the processes are studied by both thermodynamic and kinetical methods.

A thermodynamic system (or any functional subsystem of it) is specified by fixing its boundary and therefore, by determining the space domain the system (subsystem) occupies. It is also important to indicate the adequate time scale in which the behaviour of the system or some process is to be studied. This helps to reduce considerably the number of variables that should be taken into account when constructing models of real systems or events.

While studying the biological evolution, one usually finds out how its features are influenced by the intrinsic factors (genetic, in particular) and by the conditions of the environment. The general answer is unambiguous: there is influence of both sorts. However, the detailed (quantitative) answer depends on the choice of the observation time scale and on the rates at which the parameters of the biosystem and the environment vary.

The "hierarchic approach" suggested recently [2,3] allows to separate thermodynamics from kinetics and to point out the motive force of evolution processes in ontogenesis, philogenesis and at durable periods of the biological evolution. Such an approach takes into account the hierarchic structure of the biological world and uses the principles of deduction and reductionism.

The macrothermodynamic model [2-16] is based on the ideas about the origin of hierarchic structures. For instance, structures of hierarchy j are formed due to the aggregation (self-assembly) of lower-hierarchy ($j-$

1) structure elements. At each hierarchic level the substance aggregation processes are considered as partial evolutions. The model takes into account only the structure formation phenomenon, that is, it can be used for the study of the evolution of structures (or, more precisely, structural systems).

While investigating the evolution processes from the view point of thermodynamics, one should first of all specify the system under study and point out its surrounding - the thermostat (in the broad sense of this term, see Appendix 2), whose parameters are almost constant at times comparable with the life-span (evolution time) of the system. For instance, for the system represented by a cell (j), the thermostat is formed by the bio-tissue of the corresponding organ of the organism under consideration $(j+1)$. The bio-tissue is of higher hierarchy than the cell (its life-span is equal to the life-span of the organism), and it lives few dozens times longer than the cell. Note that here the systems of j-th hierarchy are denoted by j, the systems of $(j+1)$-th hierarchy as $(j+1)$ and so on.

It will be shown that the above-mentioned system (j) can be considered as quasi-closed both in the thermodynamic sense (for the case of a cell, at times of relaxation to local intermolecular equilibria) and, in part, in the kinetic sense (at times comparable with the life-span times of the cell, under the condition that the flow of substance coming into the system has constant composition, which is equivalent to the existence of a thermostat). The last statement actually means that the specific Gibbs function (or some other appropriate state function) corresponding to the formation of the system (j) from lower-hierarchy $(j-1)$ structure elements, tends to a minimum. We note that the system (j) and its thermostat (j+1) form a complete thermodynamic system, which is also quasi-closed because it is located in the thermostat of a higher hierarchic level (j+2).

To be sure, there are plenty of natural systems consisting of a system itself and its thermostat. Under real conditions, parameters of the thermostats fluctuate near their average values. The model admits these parameters to be averaged at sufficiently large times. However, this fact does not influence the tendencies in the variations of the corresponding specific Gibbs functions in the course of the evolution. The statements formulated above can be applied to all hierarchies, including populations, species, and societies. Anyway, such an assumption is justified when one studies physico-chemical transformations in ontogenesis, philogenesis, and the evolution of plants and animals at durable periods of the biological evolution. For instance, different cells (of the organs and tissues of plants and animals), which can be considered as thermodynamic systems, "live" in a thermostat - in the organism. (At corresponding times, these systems accumulate chemical substance and, hence, are partly kinetically quasi-closed with respect to the flow of substance *coming out* of them.) As to the organisms that are "evolution-distant" from each other, their environments, or thermostats, are also of essentially different nature. This fact causes considerable difficulties if the evolution tendencies of the development of such objects are to be understood. Nevertheless, there are grounds to suppose that the thermodynamic tendencies of the evolution of bio-systems, though being rather fuzzy, can manifest themselves not only at durable periods of biological evolution, but also during all the evolution of life on the Earth. The fact that there are thermodynamic tendencies in the biological evolution does not cancel the important role of kinetic and random factors in the evolutionary processes, as we have already mentioned. These two factors indeed influence the ways how "chemical and supramolecular" substances transform in the course of the evolution, but the ways of transformation are always controlled by thermodynamics. The role of thermodynamics manifests itself either in the "tendency" of processes (which follows from the second law in the form of a uni-directed "fuzzy tendency" of the entropy increase on the molecular and supramolecular levels), or, maybe, in the "irreversibility" of those processes. (Note that irreversibility should be understood as a more broad notion than the entropy increase [17,18]).

During the biological evolution, there occurs branching that manifests itself in the sequences of events justified thermodynamically but random if considered at small time intervals. Due to the branching, the "averaging" of thermodynamic parameters of systems and their thermostats acquires more and more restricted efficiency. The thermodynamic tendencies can be described as "fanning". At higher stages of the evolution, which is considered as a process of structure formation, the role of entropy factors decreases, while the possibilities of the biosystems development are mostly determined by phenomena independent of the entropy increase [9,17]. This can lead to the branching that increases the role of random events in the evolution process. It is interesting to note that hierarchic thermodynamics (macrothermo-dynamics), which studies the evolution of higher-hierarchy complicated systems, can consider separately mechanical and other processes in biosystems that are not related to thermal effects and are therefore non-entropic. It should be also stressed that K. Denbigh [18] in his philosophic view on the nature of irreversibility considers not only physical non-entropic phenomena, but also some non-entropic processes that hardly admit strict definition. These processes, probably, can be characterised as speculative (for instance, the phenomenon of onto-genesis considered independently from physiological, biochemical and physical processes).

Considerations presented above can probably lead to an opinion that the evolution is often "governed" by chance. However, this is not quite correct: the joint action of random events in a thermodynamical system should always satisfy the requirements of thermodynamics. The "fan of thermodynamics" always has a fixed direction. Things are different when the evolution leads to the appearance of a small number of objects (for instance, systems consisting of a few totalitarian superstates). In this case, the role of thermodynamics (sociological thermodynamics) turns to zero, and the behaviour of the system, which is not thermodynamic any more, becomes unpredictable, i.e., determined by purely random factors.

It should be noted once more that the thermodynamic tendencies are of statistical origin, and therefore they are unambiguous only for systems containing a sufficiently large amount of objects and at sufficiently large time intervals. Generally speaking, random behaviour of systems with small number of objects at small time intervals does not contradict to the thermodynamic tendencies at large time intervals.

Thus, here I have presented a few general principles, which, to my opinion, should form the basis of the model of biological evolution (as well as the evolution in general) taking into account the general natural laws.

The investigation "tool", if one can put it so, chosen by this model is the optimal notion (rational, constructive, and convenient for the current study) among the pairs "thermodynamics - kinetics", "organism - environment", "external factors - internal factors", "regularity - random behaviour" and so on. The choice of the concept depends on the space and time scales, which are related directly to the scale of energy needed for the formation of the material objects under study.

Using the above-mentioned concepts, one can describe various phenomena by means of models that are close to reality. However, the reality is much more complicated than our image of it.

` Another important thing to be mentioned is the well-known statement that each model or theory must be sufficiently simple and testable experimentally. One of the models of biological evolution satisfying this requirement is macrothermodynamic model. As it has been already mentioned, this model describes, beside partial evolutions (structure formation processes), the evolution of chemical substances constituting the supramolecular structures of cells. It can also consider the evolution of organisms, populations, species, other higher-hierarchy biosystems, including the human society. Note that any system of a given hierarchic level can be characterised by the value of the specific Gibbs function of structure formation,

$\overline{\widetilde{G}}_j$, which is calculated with respect to a standard level - therefore $\overline{\widetilde{G}}_j$ is often denoted as $\Delta\overline{\widetilde{G}}_j$. For the sake of clarity, we draw the reader's attention to the fact that each partial evolution (i), that is, the self-assembly of structures relating to the hierarchy j, takes place "inside" the system of structures relating to all higher hierarchic levels ($j+1$, $j+2$, ...).

Variation of $\overline{\widetilde{G}}_j$ (or variation of $\Delta\overline{\widetilde{G}}_j$) corresponding to the formation of a certain structure, which can be a supramolecular structure, a structure of populations, and so on, is a measure of the evolutionary transformation of this structure. Indeed, for the bio-tissue of some organ of the organism under study (which is a simple thermodynamic system; see Appendix 2) the evolutionary development can be characterised by the value of $\Delta\overline{\widetilde{G}}_j^{im}$ corresponding to the formation of the phase of supramolecular structures. The value of $\Delta\overline{\widetilde{G}}_j^{im}$ (or the value «$\Delta\overline{\widetilde{G}}_j^{im}+const$», which correlates with it), can be easily determined, for instance, in terms of the two states model using the method of differential scanning calorimetry. The value of $\Delta\overline{\widetilde{G}}_j^{soc}$ corresponding to the formation of some human society (a complicated thermodynamic system) can be estimated calculating the work that is to be produced in order to "build" the structure of this society. (Such an estimate could be probably made introducing some absolute hard currency.) In this last case it is necessary to use the thermodynamics of complex systems and to solve several new scientific problems [7-9].

2. EARLIER STUDIES

Twenty years have passed since the publication of papers [2,3] where the author claimed that the peculiarities of the biological evolution can be understood from the view point of a physical theory based on the achievements of classical thermodynamics – thermostatics. Using the concept of quasi-equilibrium and revealing the temporal hierarchy, which corresponds to the structural hierarchy of the biological world, the author was able to develop the macrothermodynamic model describing the evolution of biosystems. It became possible to select monohierarchic and polyhierarchic heterogeneous quasi-closed systems (subsystems) such that their behaviour could be characterised by the specific values of thermodynamic potentials tending to their extremal values within the fixed time scales.

It was postulated [2,3] that since the specific Gibbs function of supramolecular interactions corresponding to the biotissue structure formation tends to a minimum, $\overline{\widetilde{G}}^{im} \to min$ (or $\Delta\overline{\widetilde{G}}^{im} < 0$), there occurs the self-assembly of chemical structures (consisting of molecules and molecular aggregates) that are rich in energy-intensive substance. As a result, the specific Gibbs function of the chemical substance in a natural open biosystem (that is, kinetically quasi-closed system) should tend to a maximum, $\overline{\widetilde{G}}^{ch} \to max$ (or $\Delta\overline{\widetilde{G}}^{ch} > 0$).

Fig.7 shows the theoretical scheme describing variation of the chemical (*ch*) and supramolecular (*im*) components of the specific Gibbs function of the biotissue of organisms in the course of ontogenesis (*ont*) and philogenesis (*ph*). (The Gibbs function of chemical substance, like the chemical potential of a separate component, is divided in two parts - the chemical one (*ch*) and the supramolecular one (*im*). This scheme, first presented in papers [2,3], is now reliably confirmed in experiment.

Let us make a few remarks for clarity. A biological system, say, on the level of community, changes its composition in the course of the evolution - that is, molecular, supramolecular, cellular, organismal, populational composition, and so on. If we restrict the consideration to the study on the "chemical" level, we shall observe the tendency shown in Fig. 7. According to Fig.7, the chemical composition (*ch*) of the open system (which is partly kinetically quasi-closed with respect to the out-coming flow of substance) recedes from its initial state in the course of evolution (see also Fig.4 of Chapter I). Such an effect lead many scientists to the conclusion that a system recedes or "runs away" from some equilibrium state, which has usually no strict definition. This, in its turn, results in the statement that the phenomenon of life can not be explained from the view point of the second law in the framework of equilibrium thermodynamics.

To the author's opinion, the situation is completely different. The only correct statement here is that the biosystems tend to local equilibrium "conformations", i.e., to local supramolecular (*im*) equilibria corresponding to the given living conditions for the organisms. This tendency, which can be easily understood from the view point of classical thermodynamics, is the physical-chemical motive force of the evolution process. As to the changes in the chemical composition, this is a "secondary fact". Here we find a direct analogy to the effect of adsorption of some energy-intensive substance by a porous adsorbent particle placed into a solution. (Note that the particle itself is an open system.) However, in this case nobody would dare to claim that the adsorption process does not agree with the second law in the framework of the equilibrium model!

Because of the importance of the question discussed, let us make the following comment.

One should always keep in mind that the rationality of thermodynamics is caused by the fact that the second Principle can be applied to systems and processes. The thermodynamic method, which is first of all connected with the use of state functions achieving their extreme values, is effective for the study of systems that do not exchange matter with the environment. By definition, isolated and closed systems are of this type. Thermodynamics pertains its rationality also when applied to the study of processes localised in such systems. Localisation of this kind is also possible in open systems but only at appropriate times, when an open system can be considered, to a certain approximation, as isolated or closed. In this case it is more convenient to speak about «quasi-isolated» or «quasi-closed» systems.

Strictly speaking, thermodynamics (thermostatics) in general case it is not convenient to study open systems. The evolution of an open system advisable should be considered as a sequence of transitions, for instance, from a closed system with given composition to new "independent" systems of different compositions. Schematically, it looks as follows:

System 1 → *System 2* → *System 3* → ... *System n*

Each of these systems (1,2,3,...n) exists for a certain moment corresponding to the time interval when each system can be considered as closed (quasi-closed). Hence, it is evident that one should not talk about the evolution of an open system the same way as about the evolution of a thermodynamic system, which can be studied by means of standard thermodynamic methods. As we have already mentioned, it is not rational to study the evolution of such open systems in terms of "going away" from the "initial" equilibrium. It seems that in this case a more constructive approach is to investigate how the composition and structure of each new quasi-closed thermodynamic system differ from compositions and structures of similar systems that existed previously (1,2,3,...n). From the viewpoint of thermodynamics, the differences of such systems manifest themselves, for instance, in the differences between the Gibbs (Helmholtz) functions of their formation. They are denoted in this work as ΔG_j(j = 1,2,3,...n). For instance, the above-mentioned difference between the Gibbs function of supramolecular structure formation for system 1, ΔG_1^{im}, and the Gibbs function of supramolecular structure formation for system 2, ΔG_2^{im} is characterized by the value

$$\Delta\Delta G_{1\rightarrow 2} = \Delta G_2^{im} - \Delta G_1^{im}.$$

Thus, while studying the evolution of a natural open system, in the general case the term "open system" should not be identified with the notion of an individual thermodynamic system. It is better only to speak about the transformation of the initial system into new systems with different composition.

The present model of thermodynamical self-organisation (self-assembly) has been applied to the higher stages (hierarchical levels) of the biological evolution, including the evolution of populations and community [2,3].

In spite of the fact that all basic ideas presented in [2,3] were fruitful, as it became clear later, their understanding faced some difficulties. Probably, this was caused by the absence of unambiguous quantitative data that would allow one to consider the biosystems as quasi-closed thermodynamically and kinetically. Moreover, the prevailing viewpoint at that time (as well as nowadays) was that the methods of thermostatics cannot be applied to the biosystems in the process of evolution, since such systems are open at all times and considerably distant from some "environment" (the last notion had no strict definition). This viewpoint was not helpful for the understanding of the ideas presented in this book. The understanding was also hampered by the new, non-conventional terminology, which was necessary for the description of hierarchic models.

Later, the author's desire to express his ideas in a more "attractive" form, lead him to the publication of papers that contained the explanations of certain statements but were overloaded with information of, probably, minor importance. For instance, the paper [6] dealt presumably with the thermodynamics of biochemical processes, which only indirectly related to the study of the structure evolution of the biological systems.

In paper [4] the authors managed to create the thermodynamic theory of quasi-closed systems behaviour based on the Le Chatelier-Brown principle and to verify the well-known Weber-Fechner law, which was obtained earlier from empirical considerations. At present, all general ideas developed in that work turned out to be correct. However, it seems that the authors of [4] did not pay proper attention to the ways how to select quasi-closed functional systems whose behaviour was described by means of the Le Chatelier-Brown principle. (A reader might ask whether this principle could be applied to open systems.) Probably, it would be reasonable to stress the fact that the behaviour of functional systems under study (for instance, sensor systems) is certainly investigated only at time intervals where the systems to a good approximation can be considered as closed.

In some studies [8,9] simple models have been suggested that enable one to understand the physical sense of modeling the evolution of living systems.

In order to find out, in the framework of the ideas formulated above, what is the motive force at various stages of the biological evolution, let us use the following reasoning. Consider a simple sequence of cyclic synthesis and decay processes for a certain polymer, namely, polyacrylonitrile, in a model reactor. One of the zones of the reactor contains the monomer (acrylonitrile) vapour together with its water solution, where polymerisation takes place, for instance, due to the influence of light (analogue of photosynthesis; see ref. Lord Porter, Ref. to Ch. 1) [19]. The synthesised polymer is not water-solvable and is segregated as a solid phase. Gravitation (centrifugal) force moves the polymer to some other zone of the reactor. Further, due to the action of intensive radiation the polymer is depolymerised, and the resulting monomer is again fed into the gas phase of the first zone of the reactor. Then the monomer is again put into the solution, gets polymerised, and so on. If depolymerisation always lead to the formation of the monomer - acrynitrile, the cycle would be stationary and the process would go on infinitely long. However, polymerisation is accompanied by the creation of macromolecules with different molecular masses and slightly inhomogeneous macromolecules structures. Moreover, products with higher molecular masses are depolymerised, on the average, at a lower rate than the ones with lower masses. (Depolymerisation also results in the creation of new structures, including the cros-linked ones.) This leads to the accumulation of slowly decomposing substance in the depolymerisation zone. From the view point of thermodynamics, the cycle is a rough model of the "stages" of ontogenesis and the biological evolution. More specifically, formation of solid polymer is analogous to the creation of the biomass, while the destruction of the polymer corresponds to the biomass degradation.

Fig.8 shows quantitative data relating to an ideal model reactor where polyacrylonitrile degrades forming its monomer (25°C). For the left-hand curve, $\Delta G_j = f(S')$, the following points are depicted: 1 corresponds to ΔG_j ($\Delta G_{f_{298}}^0$) for one mole of the monomer in the gas phase; 2 to the same value for one mole of acrylonitrile in the water solution; 3 to one mole (per one molecular unit) of swollen solid polyacrylonitrile. The point 4 relates to "solid" polyacrylonitrile containing no monomer. (This situation is possible when all monomer contained in the particles is polymerised and no new portions of monomer can penetrate into particles since all particles are solid.) The points shown on the right-hand curve, $\Delta G_j = f(S')$,, relate to the volume densities of the Gibbs function of the structure formation, $\overline{\Delta G}_j$ ($\overline{\Delta G}_{f_{298}}^{0}$) for the following subsystems: $1'$ - acrylonitrile in the gas phase; $2'$ - acrylonitrile in solution; $3'$ - swollen polyacrylonitrile; $4'$ - solid polyacrylonitrile. The point "0" on the abscissas axis shows ΔG_j for the system "$3C+1,5H_2+0,5N_2$" (moles). Note that ΔG_j and $\overline{\Delta G}_j$ are differences of the Gibbs functions for two different states, one of them being taken as a standard, as it is usually done in thermodynamics. Clearly, from the view point of energetics the energy supply for the cycle is provided by light (radiation) and gravitation. From the view point of the formation of complex molecules and solid substance (polymer), the corresponding transformations take place in the reactor according to the laws of thermodynamics. In the usual sense, thermodynamics is the motive force of spontaneous processes. Certainly, the proper ways of transformations are determined by kinetics, but everything takes place in accordance with the requirements of thermodynamics. Further, it is clear from Fig.8 that the formation of hierarchic structures, such as the monomer, macromolecules, polymer aggregates in the solution, and particles of solid polymer, is accompanied by the "concentration" of the Gibbs function chemical component in higher-hierarchy structures. Quite similarly, one can observe concentration of the volume (mass) density of the Gibbs function $\overline{\tilde{G}}_j$ ($\overline{\tilde{G}}_{j+n}^{ch} \to max$) in polycomponent structures of different hierarchies in the course of biological evolution. This is possible owing to the effect of structure stabilisation and the existence of different scales of relaxation times for hierarchic subsystems. Higher partial evolutions [9] have higher rates than lower partial evolutions, and therefore the former remove the products of the latter from the "reaction zone". This isolation of the products of lower evolutions is equivalent to the structure stabilisation.

The initial thermodynamical system tends in the course of the evolution to the "internal" (for instance, supramolecular) equilibrium. However, new spatially selected subsystems that appear in the course of the evolution and can be considered separately, seem to be receding from some standard reference point (initial partial chemical equilibrium with the initial environment). This can be seen in Fig.8 where a simple model is considered. Note that, as we have already mentioned above, the evolution changes considerably the nature of the systems - systems of higher hierarchic levels appear that are spatially separated from the systems of lower hierarchies. One should take into account that it is probably not reasonable to consider the problem of equilibria between structures of different hierarchic levels, for instance, because such "equilibria" cannot be characterised in terms of the evolution potentials (for instance, chemical potential). For example, one cannot study the "equilibrium" between the water molecules in vapour and the particles of ice - the equilibrium is possible only between the phase of water molecules in vapour and the phase of water molecules in ice. In other words, one should not speak about equilibria between some components and the phases formed by these components. Hence, the very formulation of the problem about the biosystems receding in the course of evolution from "some undefined imaginary equilibrium state", which existed in the past, is quite problematic and has certain restrictions in each specific case. The processes (equilibria) taking place "inside" various hierarchic levels are studied by the hierarchic (structural) thermodynamics, whose basic methods are presented in [7-9, 20] (see Appendix 1).

During the recent decade, some works appeared that, to a certain extent, promoted the utilisation of the ideas formulated earlier. Starting from 1992, the author concentrated on the development of the models describing only the evolution of chemical and supramolecular structures in ontogenesis and at durable periods of the biological world development. Processing of numerous experimental data allowed to prove that the adaptation of the chemical composition of biotissues to the changes of the environment parameters has thermodynamic nature.

A more precise and clear presentation of the models suggested earlier was achieved by "splitting" the concept of thermodynamical "quasi-closeness" in two components: the "namely thermodynamical quasi-closeness" and the "kinetical quasi-closeness". This procedure, together with the study of the evolution of supramolecular structures composition based on the concept of phase equilibria existing in the systems, increased the number of scientists who considered the model to be consistent with the reality.

At present, it seems reasonable to give a more detailed description of one of the most well-studied partial evolutions of living systems - the evolution of supramolecular structures, their chemical composition and construction. It is also useful to describe some experimental results confirming that the present model can be applied to the explanation of the biological evolution.

3. THE MODEL OF A LIVING SYSTEM EVOLUTION

Let us consider some volume of biotissue (biomass) as a heterogeneous thermodynamic system consisting of *the liquid phase* - physiological substances in water solution - and *the phase of supramolecular structures* (the supramolecular "core" of biological structures), which arises as a result of the aggregation (self-assembly) of molecules, macromolecules, and supramolecular formations of different hierarchies [12, 13, 16, 21-25]. We have all reasons to assume that the phase of supramolecular structures appears due to weakly non-equilibrium phase transitions.

The simplified model of a double-phase open system suggested here describes two states relating to any local volume of the biomass. It can be applied to supramolecular formations of any type: aggregates of molecules, organells, cells, biotissues, organs, or organisms.

Local supramolecular equilibrium is achieved in all points (microvolumes) of the phase of aggregated supramolecular structures, which also includes small molecules. Therefore, the effect of self-assembly leading to the formation of the phase of supramolecular structures will be called *thermodynamic self-organisation*, in contrast to the *dynamic self-organisation*, or simply self-organisation, in the notation of I.Prigogine [26], which is observed in systems that are far from equilibrium. The existence of local equilibrium means that at times comparable with the periods of relaxation to equilibrium we deal with a set of thermodynamically closed microvolumes constituting the phase under consideration. (A microvolume is an elementary volume that is still so large that it contains a great number of particles.) Clearly, the integral value of the specific (averaged over the volume) Gibbs (or Helmholtz) function corresponding to the formation of the "averaged local conformation" of aggregated supramolecular structures, $\overline{\widetilde{G}}^{\,im}$ ($\Delta\overline{\widetilde{G}}^{\,im}<0$) achieves its minimum:

$$\overline{\widetilde{G}}^{\,im} = \frac{1}{V}\int_0^V \frac{\partial \widetilde{G}^{im}}{\partial m}(x,y,z)dxdydz \;\rightarrow\; min; \qquad min \equiv \overline{\widetilde{G}}^{\,im}_{min,1} \tag{1}$$

For the phase of supramolecular structures with constant composition
(at times of relaxation to local equilibrium)

Here V is the volume of the system, m the mass of the selected microvolumes; x, y, and z are coordinates; the symbol "—" means that we consider the specific value of G^{im}; the symbol "~" stresses that the system is heterogeneous.

The specific im-component Gibbs function of a heterogeneous system per unit mass can be represented in the form:

$$\overline{\widetilde{G}}^{\,im} = \frac{1}{M}\int_0^M \frac{\partial \widetilde{G}^{im}}{\partial m}(x,y,z)dm , \tag{1'}$$

where M is the mass of the system.

The values of m for the selected microvolumes can be rather large since the "averaged local conformations" are formed quickly. This is due to the fact that macrovolumes contain the molecules of water and other low-molecular components. The structure of the water solution determines the "mobility" of intermolecular equilibria. Note that at present nobody seem to doubt the validity of the expressions (1) and (1').

Note that our model admits averaging of the values $\Delta\overline{\overline{G}}^{im}$ and $\Delta\overline{\overline{G}}^{ch}$ with respect to the volume (mass) of any biological objects: supramolecular formations, organells, cells, biotissues, organs, organisms, populations, etc. A similar procedure of dividing the volume of a biosystem in parts exists in mozaic non-equilibrium thermodynamics: a system is divided in "(bio)chemical and (bio)physical elements" constituting it.

At times much exceeding those needed for relaxation to the local intermolecular equilibria, the bio-system is naturally open - there is a flow of chemical substances passing through it. The biosystem seems to be blown up, its volume and mass increase. The model assumes that the flow of matter *coming into the system* has constant average composition (though the flow rate fluctuates near its average value). So the nature of the substances *coming* into the system (into the phase of supramolecular structures) practically does not vary. In other words, the supramolecular phase (structure) of an organism or of some organ undergoes an evolution "against the background" of the flow of chemical substances of almost constant composition *coming into the system*. Note that here we speak only about the constant composition for the *in-coming* flow. The *out-coming* flow does change its composition. The system cannot be considered as stationary at times comparable with its life-span.

If the flow of substances coming into the system is slow enough then we can assume that the liquid phase of the biosystem is always in equilibrium (quasi-equilibrium [9]) with the flux. This provides constant average concentrations of substances *coming into the liquid phase with the flow*. (The liquid phase together with its environment can be considered as a thermostat for the phase of supramolecular structures, see Appendix 2.)

There is a strict experimental proof of the assumption that the time-averaged composition of the flow of chemical substances coming into a bio-system from the surrounding is constant. It is shown [9,10,13] that the sequences of hierarchic natural structures (arranged according to the energies or the Gibbs functions of their formation: $...\gg \Delta E^{ch,i} \gg \Delta E^{im,i} \gg \Delta E^{cel,i} \gg...$ and $\gg \Delta G^{ch,i} \gg \Delta G^{im,i} \gg \Delta G^{cel,i} \gg...$) correspond to the series of average life-times (relaxation times) of these structures in the biomass or biosystem. For instance, for a separate community of several close species one can write

$$...\ll t^m \ll t^{im} \ll t^{organell} \ll t^{cel} \ll t^{org} \ll t^{pop} \ll t^{soc} \ll... \qquad , \quad (2)$$

where t is the average existence time (life-time) for "free" metabolite molecules, supramolecular structures, organells, cells of the biotissue, and also organisms, populations, community (see Chapter I, Fig.6).

The series (2) is a geometric progression of the form $t_n \cong t_0 \beta^n$, where t_n is the average life-time for structures of the n-th hierarchy within the selected biosystem; $n = 1, 2, 3, ...$; t_0 is the standard time equal to the average life-time for the structure of the

lowest (standard) hierarchy (0) of the series (2); β is
a approximately constant for the given series. The rule
(the law) represented by the series (2) can be
formulated as follows: *Structures of lower hierarchy
(j) of the biosystem (J) exist inside the structures of
higher hierarchy (j+1) of the same biosystem (J) for
the time-period (t^j) that is much less than the life-
time (t^{j+1}) of the higher-hierarchy structures (j+1).* A
laconic formulation of the rule (2) is given in Chapter
I.

It is the low (2) that divides the time axis in
many parts and thus allows us to consider separately
the thermostat (the environment) and the system under
study j (or J) that forms, together with its thermostat
(j+1), a complete thermodynamical system [j+(j+1)]. As
we have already mentioned, this enables one to talk
about quasi-closed systems existing in the biological
world. These systems develop under the conditions of
almost constant kinetic factors determining the flow of
the corresponding substance from the *environment* (the
thermostat). To a certain extent, selection of quasi-
closed systems helps to avoid the difficulties arising
when one tries to describe the behaviour of systems of
this type (open at large times) using state functions.

Indeed, as we stressed several times, the average
life-span of separate cells belonging to an organism
(an organ), as a rule, is a few dozens times less than
the life-span of the organism itself. It is hence clear
that the organism (the organ) is a thermostat (in the
broad sense) for the cells constituting it. Its
parameters, such as T, p, concentrations of chemical
components, etc., are constant. It will be shown later
that the existence of a thermostat allows to consider a

cell (or a set of cells) as a kinetically quasi-closed system only with respect to the flow of substances coming from it. Note that since different hierarchic structures, such as organells and cells, have different life-spans, their supramolecular structures are characterised by different times of existence. However, this does not influence the thermodynamic tendencies of the systems that tend to supramolecular equilibrium in any local volume of the biomass.

It is important to stress that in the earlier papers [2-16] the model of the evolution was based not on the series (2) but on the hypothetical series of times needed for the relaxation to some imaginary equilibria between different structures of similar type within a single hierarchy (for a given volume of biomass, imaginary processes of "mixing" between different uni-type structures of the same hierarchy were considered. The relaxation series was written in the form:

$$\ldots \gg \tau^{m} \gg \tau^{in} \gg \tau^{organell} \gg \tau^{cel} \gg \tau^{organism} \gg \tau^{pop} \gg \tau^{sos} \ldots) .$$

That approach, though actually equivalent to the present one, was too complicated and therefore its understanding by the colleagues faced certain difficulties. Note that the paper (*J. Biol. System, v. 4, N 2, 1996, 239*) contains technical misprints: the signs of strong inequalities in the seria (1.1) and (1.4) should be replaced by the opposite ones.

Chemical composition of a biosystem's phase of supramolecular structures varies slowly at times comparable with the durations of adaptive processes and ontogenesis. (It also varies in the course of philogenesis and at long periods of the biological evolution.) As the biotissue gets older, the supramolecular structures become more stable thermodynamically. (Here we mean the stability of supramolecular structures but not of the chemical substance they contain.)

Selection of supramolecular structures with higher thermodynamic stability (structural phase stabilisation) is determined, first of all, by the thermodynamic factor. Indeed, it is admitted that the retaining (the retention) time (this term, as well as the model of the substance "exchange" in biosystems, is taken from chromatography) of molecules (macro-molecules) in the supramolecular phase, t_{ret}^{in}, is exponentially related to the value of the Gibbs function of supramolecular structure formation [7,9,10]:

$$t_{ret}^{in} = A \exp(-\Delta \overline{G}^{in} / RT) , \tag{3}$$

where A is a coefficient that slowly changes as the chemical composition of a biological object changes in the course of its evolution (in principle each microvolume is described by its coefficient A), R is the gas constant, and T the absolute temperature. For one of the models, A is equal to $t_{d}(\overline{V}_1 / \overline{V}_2)$, where t_{d} is the "transport time" of the flow, $\overline{V}_1$ and $\overline{V}_2$ are the specific volumes of the immovable (the fixed) and movable phases (Gerasimov, 1985, Chapter I).

The most long-retained molecules of the supramolecular phase (coming into various hierarchical structures of a biosystem from the environment or produced in the course of the biosynthesis) initiate the selection of similar molecules. This stimulates the changes in the composition and chemical nature of the phase of supramolecular structures. As we have already pointed out, these changes result from the thermodynamic factor, though it manifests itself in the kinetic form (see Equation (3)). Thus, it is presumably those molecules accumulated in the phase of supramolecular structures, whose absorption (self-assembly) is most beneficial thermodynamically. (These molecules have higher affinity to the substance forming the phase of supramolecular structures.) If there are mechanisms of matrix synthesis, such molecules have advantages for reduplication (reproduction). Due to all this, the value of the specific Gibbs function of supramolecular structure formation, $\Delta \overline{\widetilde{G}}^{im}_{min,1}$, or the value of the specific Helmholtz function of supramolecular structure formation which, for the condensed phase, practically coincides with it, grows during the evolution of the biotissue and becomes more negative. It follows that

$$\overline{\widetilde{G}}^{im} = \frac{1}{V}\int_0^V \frac{\partial \widetilde{G}^{im}}{\partial m}(x,y,z)dxdydz \rightarrow \min ; \qquad \min \equiv \overline{\widetilde{G}}^{im}_{min,2} \quad . \qquad (4)$$

For the phase of supramolecular structures of varying composition (for the times of ontogenesis, philogenesis, etc.)

The expression (4) means that the value of $\overline{\widetilde{G}}^{im}$ ($\Delta \overline{\widetilde{G}}^{im} < 0$, Eq. (1) of the present Chapter), whose minimum is achieved at small times and corresponds to a local equilibrium ($\overline{\widetilde{G}}^{im}_{min,1}$), in the course of ontogenesis (and also philogenesis and long stages of evolution) varies slowly ($\overline{\widetilde{G}}^{im} = f(t)$), tending to a still lower, more negative, value ($\overline{\widetilde{G}}^{im}_{min,2}$).

Let us note again for clarity that Equation (4) follows from the fact that the phase of supramolecular structures (or bio-tissue as a whole) is quasi-closed in relation to the *out-coming* flows of matter. This quasi-closeness leads to the accumulation of supramolecular structures (aggregates of molecules) with higher thermodynamic stability in the unit volume of the phase of supramolecular structures [9, p.90-92]. The phase of supramolecular structures must get enriched with stable superstructures for the subsystems of all hierarchic levels (cell membranes, microcorpuscles, etc.), if the flows coming into these subsystems have constant composition. In other words, at corresponding times there should exist a lot of subsystems of the type "structure - thermostat". The nature of such subsystems varies at large times but the tendency of the processes remains thermodynamic. However, certain subsystems "structure - thermostat" can transform in the course of the evolution, due to some changes in the environment, so that the gradients of the flows from "thermostats to structures" tend to zero or change the sign. Such subsystems become rudimentary and finally vanish.

Equations (1) and (4) actually determine the time axis (the kinetic parameter) for the variations (Δ) of $\overline{\widetilde{G}}^{im}_{\overline{\widetilde{G}}^{im}\to\min,1}$ and $\overline{\widetilde{G}}^{im}_{\min,1\to\min,2}$, and, consequently, also for their sum, whose value is negative:

$$\Delta\overline{\widetilde{G}}^{im}_{\overline{\widetilde{G}}^{im}\to\min,1} + \Delta\overline{\widetilde{G}}^{im}_{\min,1\to\min,2} = \Delta\overline{\widetilde{G}}^{im}_{\Sigma} < 0 \quad . \tag{5}$$

The last conclusion, however, as we have already stressed, relates to the case where the flow of substances *coming into* the biosystem has constant composition. If the composition of the in-coming matter varies in time, for instance, as a result of some physiological anomalies or dramatic changes in the environmental conditions (parameters of the thermostat), then the system can become not quasi-closed kinetically, and the variation of $\overline{\overline{G}}_{\Sigma}^{im}$ can become uncertain. A similar situation takes place for the case of aimed (artificial) selection, since such a selection is equivalent to the variation of the environmental conditions [9, pp. 132-133] and also for the case of neoteny.

Note that the characteristics of solar radiation also belong to the essential parameters of the thermostat. It is due to the constancy of these and other parameters of the selected thermostat in the lythosphere that the average flow of substances coming into the system does not vary. A considerable variation of the thermostate parameters (as has been repeatedly emphasized before) means the creation of a new thermostat - a new habitat. It is possible that the changes in the intensity and spectrum of solar radiation on the surface of the Earth lead to the "formation" of new thermostats and, as a result, to various events in the history of our planet. One can suppose that the extinction of dinosaurs was actually caused by the fact that they were not able to adapt quickly to the conditions of the new thermostat - the environment.

While stressing the role of photo-chemical processes in the evolution of living matter, we should mention the works by Lord Porter who obtained extremely significant results using equilibrium thermodynamics. Lord George Porter has investigated the transfer and storage of chemical and radiation «potential» in physico-chemical systems. The influences of several factors on the efficiency of solar-energy collection and storage were evaluated. It has been shown that practically indefinite storage of the products of chemical and photochemical change is possible, without significant additional thermodynamic losses, but two additional requirements are imposed: a relatively fast diffuse transport process and, in photochemical reactions, a long light period coupled with a small capacity per unit area of the reaction vessel. These requirements are more than adequately satisfied by the diurnal cycle and the fine lamellar structure of the photosynthetic apparatus.

Our model described here has a simple analogue - an adsorption (absorption) system, which is open and which slowly receives a flow of substance with constant composition. Inside the system, this substance undergoes phase or chemical changes [9, p.90-92]. Indeed, suppose that a homogeneous flow at the input contains a fatty acid and water as basic components, and there is microemulsion formed inside a column (reactor), so that the fatty acid concentration in the emulsion is high. Then the column will be soon "overfilled" with the fatty acid and will stop operating. (This is the manifestation of the fact that the system is partly quasi-closed kinetically.) Apparently, there is no doubt that the tendency presented in Eq. (4) is valid in this case [9, pp.71-75; 90-92; 163-165].

Equation (4) means that as the biotissue gets older, it must normally get enriched by chemical compounds with *the most negative values* of the Gibbs function of their *supramolecular structure* formation. Substances with high energetic capacity, having *less negative* or *positive* values of the Gibbs function of *chemical compounds* formation, $\overline{\Delta G}^{ch}$, - are namely of this kind, and this follows from the analogue of the Gibbs-Helmholtz equation - the correlation equation of the author who has established the dependence of $\Delta \overline{\overline{G}}_i^{im}$ on T_{m_i}; see equation (7) - and has been found out experimentally [9, 14-16, 21-25, 27-31]. These are lipids, proteins, polysacharides, nucleic acids and so on, - substances that under normal physiological conditions force water out of the volume of biotissue as it ages.

Table 1 contains the values of the specific standard Gibbs functions of the formation of several chemical compounds. These compounds are used by Nature for the creation of supramolecular structure of living beings biotissues. Clearly, the comparative energy intensity (capacity) of polymers, oligomers, and other metabolites synthesised from the substances contained in the table is much higher than the energy intensity of water.

A comparison of the energy capacities of systems of variable composition by comparing values of $\Delta \overline{G}_{f_{298}}^0$ is of a qualitative nature. In practical terms it is more expedient to compare the energy capacity (calorie value) of inherently different systems by measuring the amount of heat released as samples are burnt in a calorimetric bomb ($\Delta \acute{I}_{comb}^0$).

Similar tendencies in the variation of the composition of biotissues must be observed in philogenesis and at long periods of the biological evolution when the average chemical composition of the environment can be considered as constant. (In this case, one can assume that the biosystems are partly quasi-closed kinetically.)

The model considers the processes of self-assembly (thermodynamical self-organisation), no matter whether chemical reactions in the liquid phase take place in equilibrium or non-equilibrium regime. The mechanisms of substance synthesis and transport are inessential for the processes of supramolecular phase formation. Indeed, the aggregation processes just "use" various substances for the building of supramolecular structure of the biotissue. The model has also serious grounds to admit that the thermodynamic system under consideration is a simple one - by definition, only expansion work is produced in it. (This kind of work is rather small in condensed phases.) To be sure, this approximation becomes unjustified while studying, for instance, the evolution of a population, which is a structure of high hierarchy performing mechanical or any other work [7, 9, 12, 15, 17, 18]. (Here the role of interacting particles is played by organisms, and the study is focused on irreversible processes that are not accompanied by the entropy variation.)

Functioning of biological systems (for instance, of biotissues) is possible if these systems are "penetrable" enough for the matter, which is the building material for supramolecular structure. Besides, as it has been first pointed out in [12], there should exist not only internal, but also external forces leading to the "mixing" inside the substance - to the metabolism. This role is played by periodic fluctuations of the environment (thermostat) parameters around their mean values. We stress that these necessary periodic variations of external parameters are the inevitable "thermodynamic effect" of the environment on the evolution of biosystems. We obtain that only joint action of internal thermodynamic factors (observed inside the system) and external conditions (variations and oscillations of the environment physical parameters) determine the features of the evolution.

In the model presented here special attention is paid to the physical chemistry of supramolecular structures, which should be considered as one of the "keys" to the understanding of biological evolution. The fact that the biomass manifests a tendency of relaxation to local equilibria, while the incoming flow of chemical substances is constant, ensures that the value of $\overline{\overline{G}}^{im}$ tends to a minimum in the time scales corresponding to the life-spans of biological structures of all hierarchies.

Speaking about the evolution of any biological structures, we should stress that, according to the second law, structures of each hierarchy tend to fill all "thermodynamic space" available for them.

It can be easily proved [9] that the model does not contradict to the kinetic theory of Darwin and Wallace and in many cases pacifies the disputes around it.

A few additional comments to this section seem to be in order.

The existence of seria of type (2), as well as the existence of hierarchic structure of the biomass, allows one to point out, at proper time intervals, full thermodynamic systems. Any full system of such kind includes, first, the system of j-th level itself (for instance, microvolumes containing sufficient amounts of particles, or macrovolumes) and second, the thermostats - systems of higher, $(j+1)$th, hierarchic levels. This means that a system of j-th level can be considered as quasi-closed thermodynamically and kinetically (see

Appendix 2). For instance, a tissue (a biomass) can be represented as a set of quasi-closed microvolumes or macrovolumes. Each one of such volumes has its own thermostat - the environment whose parameters remain constant at times necessary for the relaxation to thermodynamic supramolecular equilibria (and local partial chemical equilibria) in such microvolumes.

We have already mentioned that the parameters of thermostats (say, the habitats of organisms) can vary at small times (for instance, at times much less than .the life-time of an organism or a tissue). Such fluctuations are caused by the changes in the environment parameters (temperature, pressure, chemical composition, etc.). If these fluctuations stay within the adaptive zone, the biosystem (say, a biotissue or an organism) develops «normally», i.e., biochemical processes, chemical and supramolecular composition of the biosystem (biotissue) can «follow» the changes in the environment parameters.

Thus, a biotissue (a system of hierarchic level $(j+1)$) can be considered as a set of subsystems of lower hierarchic level (local volumes of this tissue, j). The tissue (the organism) is surrounded by the habitat of the organism. This habitat of the organism can be considered as a thermostat with respect to all systems of lower hierarchic levels - the tissue and its subsystems. Parameters of such a thermostat are varying, oscillating («floating»). At life-times of the tissue subsystems and the tissue itself, averaging of the thermostat parameters is observed. Development of the tissue and its subsystems takes place against the background of averaged thermostat parameters. This development (evolution) has internal thermodynamical tendency, which manifests itself in the variation of the chemical composition of local tissue volumes, the tissue, and the whole organism. This is accompanied by the selection of supramolecular structures with most high thermodynamic stability. As a result, $\overline{\overline{\tilde{G}}}^{im}_{j,j+1,\ldots}$ tends to a minimum (the indices j, $j+1$ mean that $\overline{\overline{\tilde{G}}}^{im}$ can be averaged with respect to any local volume of tissue, the whole volume of tissue, etc.).

Our model describes the evolutionary development of biotissue (organism or biomass) with an account for fluctuations of $\overline{\overline{\tilde{G}}}^{im}$ caused by the fluctuations in the parameters of the thermostats. It can be called *the evolution model of mosaic hierarchic structures* or *the model of the evolution of hierarchic structures with «floating» parameters*.

Evolution of the chemical and supramolecular composition of the substance in a microvolume of a tissue (a biomass) is similar to the evolution of the substance composition in a model equilibrium chromatographic column. In such a column, kept under natural living conditions of organisms (when there are fluctuations of temperature, atmospheric pressure, and other physicochemical characteristics of the environment), chemical reactions are similar to biochemical processes in microvolumes of the biomass. Besides, the composition of chemical substances coming into the model column also varies (oscillates) with respect to some «average composition», which is analogous to the «composition fluctuations» of food consumed by the organism. It is easy to show theoretically and experimentally that at sufficiently large times, such a model chromatographic column would accumulate (since it is quasi-closed kinetically, see

Appendix 2) supramolecular structures with high thermodynamic stability. These structures should consist of chemical substances and supramolecular structures with high affinity to the sorbent filling the column. Note that at small times, thermodynamic stability of supramolecular structures can decrease slightly, due to the fluctuations of the environmental physico-chemical parameters and the composition of the flow of substance coming into the column. This causes fluctuations of the local values $\overline{\overline{\tilde{G}}}_j^{im}$ characterising supramolecular structures of the column (the index j relates to a local volume of the column). However, at comparatively large times, under the conditions imposed in the model, local values of $\overline{\overline{\tilde{G}}}_j^{im}$ tend to relative minima.

A real biotissue (biomass) can be represented by a large number of model chromatographic columns kept under local supramolecular equilibria. The macrovolume of the whole tissue (as well as any volume of the biomass) can be also considered as a model chromatographic column consisting of many microcolumns distributed locally over the volume. In this case, the averaging $\overline{\overline{\tilde{G}}}_j^{im}$ is done over the corresponding macrovolume (the index j relates to the macrovolume of the macrocolumn itself). The values $\Delta\overline{\overline{\tilde{G}}}_j^{im}(i)$ of the Gibbs function of the formation of i-th type supramolecular structures (lipid suprastructure, proteins, nucleic acids, etc.) can be determined experimentally, for instance, by the DSC method.

4. APPLICABILITY OF THE MODEL TO REAL OBJECTS

According to the model presented above, a biosystem – a tissue, an organ, etc. – under normal physiological conditions expands in the course of ontogenesis. According to Eqs (4) and (5), the system is enriched by energy-intensive chemical substances, which oust water from the biotissue.

Indeed, there has been obtained various material concerning the changes in the chemical composition of organisms (of their organs and tissues) in the course of ontogenesis [32,33] philogenesis and biological evolution in general [9, 10, 13]. (Several examples of the composition variation have been presented in Chapter I.)

A typical example of the variation of chemical compound "water – organic substances" in the brains of different animals, depending on relative levels of their evolutionary development, γ^j, is shown in Fig.9. (It is assumed that γ^j is a linear function of the concentration of water in the brain tissue; the values of γ^j for human and frog are taken to be 1.0 and 0.5, respectively.) The data on chemical composition are taken from reference literature. Comparatively large error bars corresponding to the values of m^j for the case of human (point 1 in Fig.9) and for other cases as well (points 2-14 in Fig.9) are caused by the variation of the brain tissue chemical composition of animals in the course of ontogenesis. (The error bars are shown not for all points.) Note that the data in the reference literature, as a rule, are presented without mentioning the age of the animals. The values of γ^j

for the brain of different animals correspond to our intuitive estimate of their "mental development". However, if one compares precise data corresponding to different animals at equal stages of ontogenesis, the presented "mental development" series can be reorganised considerably.

It follows from Fig.9 that in the course of ontogenesis the brain of an animal, as we have already mentioned, gets enriched by fats (lipids), proteins, and other energy-intensive compounds. Similar dependencies should be observed for other organs and tissues of living beings. Certainly, such tendencies exist for all the organs together, i.e., for the whole organism, as well as for the population and so on.

Now our aim is to show that the discussed variation of chemical composition results from the tendency of a biosystem to achieve a series of new supramolecular equilibrium in the course of the evolution. In other words, the validity of Eqs (4) and (5) is to be proved on the experimental basis.

One of the possible ways to prove the equations is to find some physical-chemical parameter describing the phase of supramolecular structure of a bio-tissue (or separate elements of this phase). This parameter would allow to judge about the value of ΔG_{Σ}^{im} of this phase formation (it characterises thermal stability of structures) and also about the chemical energy capacity of substances contained in it. It has been shown [9, 14] that the melting temperature T_m of supramolecular structures of the corresponding condensed phase can be chosen as such parameter.

For closed systems, the dependence of the Gibbs function variation on temperature T is determined by the Gibbs-Helmholtz equation:

$$\left[\frac{\partial(\Delta G/T)}{\partial(1/T)}\right]_{p} = \Delta H \quad , \qquad (6)$$

where ΔH is the enthalpy variation during the process, T and p are temperature and pressure.

An equation analogous to the Gibbs-Helmholtz equation can be also written for systems whose composition is not identical. (Note that T and p are constant.) This assumption is precise only for substances (i) that have the same values of thermodynamic parameters. Nevertheless, it proved reasonable for a certain range of the variation of thermodynamic characteristics. Indeed, the experimental data revealed the correlation (predicted in [9]) between $\Delta \overline{G}_{i}^{in}$ - the specific Gibbs function of the structure formation - and $\Delta T=T_{m}-298$, 298 being the standard temperature, K. The correlation has been obtained for non-equilibrium phase transitions "overcooled liquid - solid" for a wide range of organic compounds with T_{m_i} varying within a broad temperature interval. (The index m_i relates to the melting point of the i-th substance.)

Several results of calculation by means of Eq. 7 (p.73) relating to fatty acids are shown in Fig.10. One can see that natural substances with high T_{m_i} (i.e., with high ΔT), which have been studied by us, indeed have a tendency to form thermodynamically stable phases - solid structures (with more negative values of $\Delta \overline{G}_{i}^{in}$, see Eq. 7 (p.73), since condensation (crystallisation) processes of pure substances can be considered as enthalpy-controlled processes, from the viewpoint of the summarised effect. If more precise calculations are necessary, corrections for the specific heat variation of the substances at their melting, $\Delta \overline{C}_p$, are to be made, and so on. Note that the chemical energy capacity of a series of similar natural compounds ($\overline{G}_i^{ch}$) grows as their T_{m_i} increases ([27, 28], see also Table 1). This verifies the predicted behaviour of $\overline{G}_i^{ch}$ in the course of ontogenesis and evolution in general (Fig.7).

In the case of entropy-controlled processes, such as aggregation of cells during the self-assembly of cellular structures, denaturation of some proteins, etc., the rise in the thermal stability of the structures is accompanied by a drop of T_{m_i}. However, apparently, the relative share of such entropy-controlled processes is rather small during the structure formation in some biotissues, and also some organells.

Thus, the data of Fig.10 are in agreement with the conclusion that it is thermodynamically beneficial for living organisms to accumulate substances with high chemical energy capacity, which oust water from the biotissue.

Let us make some comments. In fact, the author has studied the correlation between the values of $\Delta \widetilde{\overline{G}}_i^{im}$ (25°C) calculated for individual closed systems (substances), and T_{m_i}. The existence of such correlations enables one to estimate the differences between the values of the Gibbs function of the structure formation ($\Delta \widetilde{\overline{G}}$) of various systems (for instance, supramolecular ones). These differences are characterised by the values $\Delta\Delta\overline{\widetilde{G}}_i^{im}$ calculated for constant temperature (T_0). The values of $\Delta\Delta\overline{\widetilde{G}}_i^{im}$ are determined by the differences in the chemical (supramolecular) composition of the systems under study ($\Delta\Delta\overline{\widetilde{G}}_i^{im} = \Delta\overline{\widetilde{G}}_{system,j}^{im} - \Delta\overline{\widetilde{G}}_{system,i}^{im}$). *The Gibbs-Helmholtz equation is* given by the relation $\Delta G = f(T)$ for a closed system. As to the *correlation equations suggested by the author*, they are given by relations of the type $\Delta G_i = f(T_{m_i})$ or $\Delta G_i = f(T_{b_i})$, and so on. The Gibbs-Helmholtz equation is a general strict thermodynamic equation, whereas the correlation equations written by the author describe particular cases. They are well satisfied only within a limited temperature interval for uniform systems (substances) of variable (different) composition. Thus, despite the limited scope of the correlations considered by the author, they can be used for predicting the thermodynamical tendency of the open systems evolution. The equation of the type $\Delta\overline{\widetilde{G}}_i^{im} = \Delta\overline{\widetilde{H}}_{m_i}^{im}\dfrac{T_{m_i} - T_0}{T_{m_i}}$, suggested by the author, where $\Delta\overline{\widetilde{H}}_{m_i}^{im}$ is the specific enthalpy variation due to crystallisation, T_{m_i} are variables and T_0 is a constant, is *an analogue of the Gibbs-Helmholtz equation* [9, 14-16]. The same equation considered for constant $\Delta\overline{\widetilde{H}}_{m_i}^{im}$, T_{m_i} and variable T_0 is the approximate *Gibbs-Helmholtz equation*.

Another obvious fact confirming the model is the adaptive, thermodynamically directed variation in the composition of fatty acids participating in the synthesis of fats [13, 30]. The data show that the cells of microorganisms, plants and animals change the composition of fatty acids (fats) when the ambient temperature changes. If the body temperature of an organism is below the optimum value, the content of unsaturated fatty acids (fatty acid residuals) increases. (Compared to the saturated and other unsaturated acids, those acids have higher melting temperature, T_{m_i}.) This increase is easily detectable from the growing iodine number of corresponding fatty acid fractions. When the temperature is above the optimum value, the cells have a greater proportion of saturated acids characterised by a higher melting temperature. Results of numerous studies are presented in papers [16, 34] and also in reference literature.

Simple quantitative considerations show that the observed variation in the chemical composition of fatty acids (fats) of biotissues due to the temperature variations in the course of ontogenesis is thermodynamically advantageously (is justify).

According to the thermodynamic model of adaptation, a flow of matter with constant composition comes into the unit volume V of a fatty tissue (or of a cell) from the thermostat - a liquid phase or the surrounding, - in which the concentration of fatty acids (fats) is maintained constant. This volume is similar to a chromatograph column that adsorbes the in-coming substances. It is convenient to use the following evident relation for calculations:

$$c_s \, / \, c_l = \exp(-\Delta \overline{G}_i^{\,in} \, / \, RT),$$

where c_s and c_l are the fatty acid concentrations in the volume V in the solid aggregated (adsorbed) state and in the liquid phase, respectively; $\Delta \overline{G}_i^{\,in}$ is the specific Gibbs function variation at the adsorption (self-assembly) of fatty acids.

Assuming that $\Delta \overline{G}_i^{im}$ are close to the values of $\Delta \overline{G}^{im}$ for the crystallisation of individual acids (Table 2), let us perform simple calculations:

$$(c_s / c_l)_{Pal,25°C} = \exp[1428 / (1.987 \cdot 298.2)] = 11.1 \quad , \quad (c_s / c_l)_{Pal,5°C} = 54.1 ;$$

$$(c_s / c_l)_{El,25°C} = 2.61 \quad , \quad (c_s / c_l)_{El,5°C} = 8.08 ;$$

$$(c_s / c_l)_{Ol\ \beta,25°C} = 0.69 \quad , \quad (c_s / c_l)_{Ol\ \beta,5°C} = 1.66 .$$

For instance, for palmitic acid at 25°C $c_s / c_l = 111$. This means that if a mole of this acid comes from the thermostat into a unit volume of a fat-tissue cell, only 0.917 mole will go into the solid phase, according to the model; 0.083 mole will remain in the liquid phase. (For convenience, the calculation is done for one mole, though the actual amount of substance in the volume V is much smaller.) At 5°C, 0.982 mole will go into the solid state, 0.018 mole remaining in the liquid phase. Thus, the amount of palmitic acid in the solid phase grows $\lambda_{Pal(25°C \to 5°C)} = 0.982 / 0.917 = 1.07$ times as the temperature drops from 25°C to 5°C.

For elaidic acid at 25°C, 0.723 mole will go into the solid phase, while 0.277 mole will remain in the liquid phase; at 5°C, there will be 0.890 mole in the solid phase and 0.110 mole in the liquid phase. Hence, $\lambda_{El(25°C \to 5°C)} = 0.890 / 0.723 = 1.23$.

For oleic acid (β) at 25°C, 0.408 mole will go into the solid phase, while 0.592 mole will remain in the liquid phase; at 5°C, there will be 0.624 mole in the solid phase and 0.375 mole in the liquid phase. Hence, $\lambda_{Ol\beta(25°C \to 5°C)} = 0.624 / 0.408 = 1.53$.

Thus, according to the model calculation, when the temperature of a biotissue changes from 25°C to 5°C, the amount of palmitic, elaidic, and oleic (β) acids grows 1.07, 1.23, and 1.53 times, respectively. Though there are several effects not taken into account by our model, the last statement is in full agreement with the above-mentioned facts indicating that a decrease in the body temperature of an animal is accompanied by the growth in the relative amount of unsaturated acids (substances with relatively low T_m) and the drop in the relative amount of saturated acids.

Note that, when estimating $\overline{\Delta G}_i^{im}$ (Fig.10, Table 2) we used the approximate equation valid for a rather wide range of temperatures:

$$\Delta G_i^{im} = (\Delta H_{m_i}^{im} / T_{m_i}) \cdot (T_{m_i} - T_0) = \Delta S_{m_i}^{im} \Delta T = -\Delta S_{m_i} \Delta T \quad , \quad (7)$$

where $T_0 = 298\text{K}$, and the superscript im shows that we consider condensation of the substance. However, this does not change the general tendency of the effect: though at melting the corrections for the heat capacity are essential (about 3-4%), they are practically equal for all substances under study.

Moreover, we compare only the values of $\overline{\Delta G}^{im}$ for fatty acids but not for the fat tissue itself. Still, this does not change the general situation either, since there are well-known strict correlations between the thermodynamic characteristics (T_m and ΔH_m) of fatty acids and fats containing the corresponding fatty-acid residuals.

Therefore, model calculations carried out with relatively high accuracy confirm that the macrothermodynamic model can be applied to the evolution of fatty tissues of organisms. In its framework, one can explain the variation in the chemical composition of fatty acids (fats) during the adaptation of organisms to the temperature conditions of the environment.

A good illustration to these statements are also classical data by Fraenkel and Hoppe describing the iodine number of phosphatides (lipids) of fly as a function of their development temperature [34]. At higher environmental temperatures, the amount of unsaturated acids drops (Fig.5, Chapter I). These data are in agreement with model calculations (as the temperature decreases by 10°C, the iodine number rises 1.23±0.03 times).

Several results relating to the variations in the chemical composition of proteins and nucleic acids in the course of the evolution of organisms are discussed in [9, 16, 21, 22, 35]. However, there is still a lack of reliable information unambigously confirming the thermodynamic nature of changes in the composition of these natural polymers. This is well illustrated by the growth of the melting temperature of chromatin in the course of ontogenesis. This growth is believed to indicate definitely that the evolutionary ageing of chromatin in the ontogenesis has thermodynamic nature [9, p.161-163]. New results [35] make it possible to conclude that the evolutionary optimisation of the RNA structure is determined not only by the thermodynamic stability of its secondary structure, but also by that of its tertiary structure. This explains why not only the sequences containing GC pairs are selected in the course of the evolution (which is most beneficial thermodynamically for the secondary structure formation). In the course of the evolution, sequences including AU pairs are also selected. We obtain (according to the theoretical predictions) that the thermodynamics of tertiary and higher supramolecular structures influences the chemical composition and structure of the RNA . Therefore, according to the author's view point, selection of natural (AUGC) sequences is most advantageous macrothermodynamically. P.Shuster [35] considers these sequences as the most stable ones with respect to mutations.

By means of calculations based on the existing experimental data, one can unambigously verify that the development of the biotissue of animals (as well as the development of separate supramolecular structures of the biotissue containing lipids, proteins, nucleic acids and so on) in the course of ontogenesis has a thermodynamic tendency. For instance, by means of differential scanning colorimetry the relation was studied between the age of the collagen tissue of a rat's tail tendon and the temperature and heat of its denaturation [36]. After the study of about a hundred samples, it was found out that as the age of the tissue varies from 2 weeks to 2 years, its denaturation (melting) temperature, T_{m_i}, increases by 6°C - approximately from 58°C to 64.5°C. According to our estimation, the enthalpy variation of denaturation increases in this aging process from 6.0 *cal/g* to 7.6 *cal/g*. The thermal capacity variation corresponding to the transition of the biotissue from the native state to the denaturated one is $\Delta \overline{\tilde{C}}_p = 0.096 cal / \mathrm{K} \cdot g$ ($T_1 \to T_2$; $T_2 > T_1$) .

Using the data presented, one can easily, with the help of the Gibbs-Helmholz equation, which takes into account the heat capacity variation at the phase transition, calculate with considerable accuracy (not accounting for the heat capacity variation with the increase of temperature) the Gibbs function variation corresponding to the supramolecular structure formation for standard (reference) temperature, T_0 ($T_{m_i} \to T_0$; $T_{m_i} > T_0$) :

$$\Delta \overline{\widetilde{G}}_{\Sigma}^{im} = \Delta \overline{\widetilde{H}}_{m_i} \frac{T_0 - T_{m_i}}{T_{m_i}} + \int_{T_{m_i}}^{T_0} \Delta \overline{\widetilde{C}}_p \, dT - T_0 \int_{T_{m_i}}^{T_0} \frac{\Delta \overline{\widetilde{C}}_p}{T} \, dT \quad , \qquad (8)$$

where $\Delta \overline{\widetilde{H}}_{m_i}$ is specific enthalpy (variation of the specific enthalpy) of melting – denaturation ($\Delta \overline{\widetilde{H}}_{m_i} = -\Delta \overline{\widetilde{H}}_{m_i}^{im}$) .

Assuming $T_0 = 298.2K$ (25°C) for the tissue of, say, a 2-year-old rat, we obtain: $\Delta \overline{\widetilde{G}}_{\Sigma,2y}^{im} = -0.66 \frac{cal}{g} \cong -2.8 \frac{J}{g}$. For the tissue of a two-week-old animal: $\Delta \overline{\widetilde{G}}_{\Sigma,2w}^{im} = -0.43 \frac{cal}{g} \cong -1.8 \frac{J}{g}$.

Calculations carried out for different stages of ontogenesis show that indeed, in accordance with the theory, the value of $\overline{\widetilde{G}}^{im}$ for the intact collagen of a rat's tail tendon tends to a minimum as the rat gets older. In the case under consideration it varies by value $\Delta \Delta \overline{\widetilde{G}}_{\Sigma}^{im} = \Delta \overline{\widetilde{G}}_{\Sigma,2y}^{im} - \Delta \overline{\widetilde{G}}_{\Sigma,2w}^{im} = -0.66 cal / g - (-0.43 cal / g) = -0.23 cal / g$. Note that an estimate according to the approximate equation (7), $\Delta \overline{\widetilde{G}}_i^{im} = \Delta \overline{\widetilde{H}}_{m_i}^{im} \frac{T_{m_i} - T_0}{T_{m_i}} = \Delta \overline{\widetilde{H}}_{m_i} \frac{T_0 - T_{m_i}}{T_{m_i}}$, yields the value $\Delta \Delta \overline{\widetilde{G}}_{\Sigma}^{im} (25° C) = -0.28 cal / g$.

One can choose as a standard temperature (T_0) the temperature of the body (tissues, cells, etc.) of the animal under study. However, this certainly does not change the general view of the phenomenon. For instance, for the studied case of the rat's body temperature equal to 38.1°C, calculations with the help of Eq. (7) give the value $\Delta\Delta\overline{\widetilde{G}}_{\Sigma}^{im}(38.1°C) = -0.22 cal/g$.

Variations of $\Delta\overline{\widetilde{G}}_{\Sigma}^{im}$ (and also of $\Delta\overline{\widetilde{G}}^{ch}$, see Fig.7) at the aging of the collagen tissue are in good agreement with the changes in the amount of water, whose concentration varies within the range 78–58% of the weight [37].

Note that the variation of the values $\Delta\overline{\widetilde{G}}_{\Sigma}^{im}$ in the course of ontogenesis connected not only with the formation of supramolecular structure with higher stability (caused by intermolecular contacts) but also with the appearance of chemical «cross-links» in the structure (see, for instance, Kanungo, reference to Chapter I).

The absolute values of $\Delta\overline{\widetilde{G}}_{\Sigma}^{im}$ (25°Ñ) obtained here for the collagen tissue are smaller than the analogous values for the condensation (hardening) of some chemical compounds from the initial over-cooled state [15, 28]. This contradicts neither the physical picture of the life phenomenon, nor the Aleksandrov semi-stable state principle and other well-known facts [17, 18, 38, 39].

There are certain results of systematic studies published during the recent years that relate to the melting (denaturation) of supramolecular structures belonging to the tissues of different animals. For instance, the works by J.Lepock and co-authors contain high-precision results obtained by measuring the heat absorption of tissues under heating (the DSC-profiles).

Fig.11 shows the DSC-profiles for two cultures of cells [42]. This figure illustrates how informative is the method. Measuring the DSC profiles for tissues in the course of ontogenesis and applying the known models, such as the model of two states, one can estimate the variation of $\Delta\overline{\widetilde{G}}^{im}$ for different types of supramolecular structures and for different tissues during the process of ageing. There are grounds to suppose that using this method, one can make some conclusions about the changes in the environment (liquid phase) of organells and cells during their regeneration. For instance, it will be probably possible to find a thermodynamic tendency in the "rejuvenating" of cancer tissue. The "rejuvenation" can be caused, for instance, by the variation of the penetrability of cell membranes, which, in its turn, should lead to the variation of the substances coming into the cells and out of them. Apparently, the development of cancer is caused not only by the transformation of genes (chromatin), but also by the variation of the environment, i.e., the thermostat.

Papers [44-46] contain important information on the evolution of the structure of proteins in the course of philogenesis (at long periods of the evolution). These results agree with the concept of the thermodynamic tendency of biological evolution.

The following fact should be stressed once more. The amount of water, m_{H_2O}, in the biotissues of living beings (j), as well as the amount of organic (and inorganic) matter, is related to the specific Gibbs functions of the formation of supramolecular and chemical structures, $\Delta\overline{\widetilde{G}}_j^{im}$ and $\Delta\overline{\widetilde{G}}_j^{ch}$. With these specific values given at least for a single moment of ontogenesis and with the variation of $\overline{m}_{H_2O}$ given for the biological object during the whole process of ontogenesis, one can easily calculate the dependencies $\Delta\overline{\widetilde{G}}_{ont}^{im} = f_2(t_{ont})$ and $\Delta\overline{\widetilde{G}}_{ont}^{ch} = f_1(t_{ont})$, similar to those depicted in Fig.7. (The changes that occur during ontogenesis in the substances constituting the biological object are not taken into account.) The data presented in Fig.1 are also in full agreement with the experimental facts.

An account for the changes in the chemical composition of a biosystem taking place during ontogenesis and philogenesis, as well as direct DSC measurements, would give corrections to the correlations obtained via the measurement of water (organic substances) content in the biological material. However, these corrections do not change the form of the dependencies (Fig.7). In the presence of a thermostat, i.e., when the system undergoes only evolutionary development and there are no revolutionary changes [9], the function $\overline{\widetilde{G}}^{im}$ of the system always tends to a minimum.

Generally speaking, there are many problems connected with the accurate estimation of $\Delta\overline{\widetilde{G}}^{im}$ of the supramolecular structure for a biological tissue at standard temperature. For instance, the DSC method can be used for estimating $\Delta\overline{\widetilde{G}}^{im}$ only for the case of the transition: *initial living structure* $\rightarrow$ *denaturated structure*. It should be noted that the conformation of molecules in the denaturated structure differs essentially from the corresponding conformations in an ideal solution. At the same time, there are certain grounds to suppose that the DSC method can provide quite satisfactory estimations for the values of $\Delta\Delta\overline{\overline{G}}_\Sigma^{im}$ in the course of ontogenesis, since denatured structures have rather close conformations for different stages of ontogenesis.

Probably, the accuracy of calculating $\Delta\overline{\widetilde{G}}^{im}$ for certain biological objects can be improved by using the data on their heats of burning or heats of solving.

There are plenty of examples analogous to the ones considered here [9, 32, 33]. One can definitely state that our thermodynamic model of ontogenesis, philogenesis, and the biological evolution in general is based on the reliable experimental material.

As we have already mentioned, the model suggested here can be also applied to the adaptive phenomena (for instance, to the adaptation of the composition of fats in biotissues to the temperature of the habitat). It allows some predictions to be formulated. For instance, it follows from the theory that the biomass (the biotissue) of living beings should contain, due to the adaptation to the external conditions, a larger proportion of heavy water (D_2O) than the environment. At present, there exist only indirect proofs of this statement. Indeed, the DNA denaturation in a water (H_2O) solution and in the heavy water (D_2O) solution is characterised by the following denaturation parameters [50]:

Solvent H_2O: $T_m = 76.9°C$, $\Delta H_m = 54.7 \pm 5\dfrac{J}{g}$, $\Delta C_p = 0.60 \pm 0.1\dfrac{J}{K \cdot g}$

Solvent D_2O: $T_m = 79.3°C$, $\Delta H_m = 70.0 \pm 7\dfrac{J}{g}$, $\Delta C_p = 0.70 \pm 0.1\dfrac{J}{K \cdot g}$.

One can see that the DNA denaturation is observed in the heavy water at higher temperatures than in the water. Hence, it follows from the equations (7) and (8) that formation of the DNA superstructure is thermodynamically more beneficial if the process involves D_2O. Indeed, calculations lead to the following result:

$$\Delta\Delta\overline{\widetilde{G}}_{\Sigma}^{im} = \Delta\overline{\widetilde{G}}_{\Sigma,D_2O}^{im} - \Delta\overline{\widetilde{G}}_{\Sigma,H_2O}^{im} = -7.83\frac{J}{g} - (-5.67\frac{J}{g}) = -2.16\frac{J}{g} = -0.52\frac{cal}{g}.$$

Thus, the DNA structure is comparatively more stable from the view point of thermodynamics if "framed" by heavy water. Hence, such a structure is preferable.

Our model of the evolution of structures allows to make a general conclusion concerning the tendency in the evolution of biochemical processes in the course of ontogenesis and in the biological evolution as a whole. For instance, selection of supramolecular structures that are most stable thermodynamically should necessarily lead to the shift of the thermodynamic equilibria towards the formation of substances participating in the self-assembly of these stable structures. In other words, due to the structure stabilisation of the products of biosynthesis, the most preferable to be synthesised are those substances that participate in the formation of the most stable supramolecular structures. To be sure, this tendency can be easily revealed under the condition that the environment does not vary (there is a corresponding thermostat). This situation resembles polycondensation taking place at the boundary between two phases. Since polymer is withdrawn from the system, in this case a reaction takes place that is thermodynamically not beneficial in an aqueous medium.

5. THERMODYNAMIC MODEL OF LIVING BEINGS' BEHAVIOUR

Applying thermodynamic methods to the quantitative study of quasi-closed biological systems, one can develop a theory based on physical grounds that would describe their behaviour on all levels, including the social and ecological ones. Such a phenomenological theory was suggested quite recently [4,9], though the approach to the problem has been known for a long time. Utilisation of Le Chatelier-Brown principle for the study of chemical, biochemical, and many other processes was extremely effective during all this century [51-57].

When the principle of Le Chatelier - Brown is applied to equilibrium (quasi-equilibrium) systems of any nature and hierarchy, it is usually called the principle of least compulsion. It is often formulated as follows: the behaviour reaction of a closed system is aimed at compensating the external perturbation. The principle is convenient, for it allows to determine the tendencies in the behaviour of thermodynamic equilibrium without analysing in detail the conditions of equilibrium.

P.Ehrenfest pointed out that any mathematical form of the principle should use both extensive and intensive parameters.

Let the state of a system be determined by the coordinates x_1, x_2, and the corresponding forces X_1, X_2. Then the differential of a certain function of state, Y, is

$$dY = X_1 dx_1 + X_2 dx_2 \qquad\qquad (9)$$

where $\quad (\partial X_1 / \partial x_2)_{x_1} = (\partial X_2 / \partial x_1)_{x_2} \quad$ and $\quad (\partial x_1 / \partial X_2)_{X_1} = (\partial x_2 / \partial X_1)_{X_2}$.

Furthermore, one easily obtains the inequality

$$(\partial x_1 / \partial X_1)_{x_2} < (\partial x_1 / \partial X_1)_{X_2} \ , \qquad\qquad (10)$$

which is the Le Chatelier-Brown principle in its mathematical form [51, 52].

The Le Chatelier-Brown principle states that in each new state, variation of the parameter x_1 is reduced due to the external force, i.e., the inequality (10) is satisfied. For instance, if some amount of heat is passed to a mixture of solid and liquid phases in equilibrium, the melting of the solid prevents the temperature increase in the system. It is commonly known that the increase of temperature during equilibrium endothermic reactions shifts the equilibrium towards the final products, while during exothermic ones the equilibrium shifts in the opposite direction. When a conductor moves in the magnetic field, there is induction current generated in it; the action of the magnetic field on the current slows down the motion of the conductor. If a magnet is put close to a conductor, the current induced in the conductor pushes the magnet away, and vice versa. The last statement expresses the Lentz rule. Another well-known example is the Peltier effect: when an electric current passes through a solder joint of two metals, the joint changes its temperature, so that the arising thermoelectric current is opposite to the initial current. The Le Chatelier-Brown principle has been applied to economy, sociology and other sciences. However, one should take into account that this principle can be applied only to closed systems, whose initial state, to a certain extent, is stable.

Applying the principle of least compulsion to model (quasi-closed) biosystems, one can write [4]:

$$\ln K_i^j = -\frac{1}{\theta} X_{\alpha_i} \Delta x_{\alpha_i}^j + const \quad , \tag{11}$$

where K_i^j is the constant characterising some current equilibrium, i, where "reaction elements" j take part; θ is constant (for physical-chemical processes, $\theta = RT$); X_{α_i} is the α_i-th generalised force (intensity factor); $\Delta x_{\alpha_i}^j$ is variation of the generalised coordinate. The product $X_{\alpha_i} \Delta x_{\alpha_i}^j$ is the work performed by the external force, or the energy fed into the system. (The sign of the work is conventional; in the present study, the work of the external force over the system is taken to be positive.)

It is often difficult to determine the amount of energy fed into a system. However, this amount is related to the energy $\Delta \Phi_\alpha$ "irradiated" by the source of the external force via the following equation:

$$X_{\alpha_i} \Delta x_{\alpha_i}^j = \varphi_{\alpha_i}^j \Delta \Phi_\alpha \quad , \tag{12}$$

where $\varphi_{\alpha_i}^j$ is the efficiency of the energy transfer, which depends on the type of the system and other factors.

It follows from the equations (11) and (12) that

$$\ln K_i^j = -\varphi_{\alpha_i}^j \Delta \Phi_\alpha / \theta + const . \tag{13}$$

The analysis of Eq. (13) shows that for the perturbation varying within certain physiological limits, i.e., when the behaviour reaction of the biosystem is able to maintain the quasi-equilibrium state, $\ln K_i^j$ depends linearly on $\Delta \Phi_\alpha$.

One can apply the minimal compulsion principle to populations and societies and write it as

$$\ln K_i^j = k\Delta\Phi_\alpha + const . \tag{14}$$

This equation should probably be treated as semi-empirical, since the physical sense of the coefficient k is not quite clear in several cases.

A lot of examples in various domains of biological and social science confirm the thermodynamic theory of behaviour. Most evident and convincing is the behaviour of animals at various physical influence over their organisms.

To check the relations analogous to (11)-(14) experimentally, one needs a thorough analysis of the effect under study. First of all, it is necessary to decide whether the intensive or extensive parameter will be chosen as the variable. This is extremely important, since the relations can be easily verified under the condition that one of the parameters is constant: either X_{α_i} or $x_{\alpha_i}^j$. Further, one should try to find out which quasi-equilibrium is related to the constant K_i^j. The estimation of K_i^j should be not difficult for homeostatic processes, which are connected with the chemical effects in the organism. Indeed, in this case, chemical or intermolecular interactions are considered. Sometimes, instead of K_i^j one can use some correlation parameter connected with it via a simple functional dependence. For instance, instead of the constant K_i^j, one can sometimes use the mass m_s of chemical superstructures or the mass m_m of some sensor-active molecules. Taking this into account, we can rewrite Eq. (14) in the form

$$\ln m_s = \beta_{\alpha_i}^j X_{\alpha_i} \Delta x_{\alpha_i}^j + const , \tag{15}$$

where $\beta_{\alpha_i}^{j}$ is constant at constant temperature.

The simplified approach presented above can be valid for the study of the biosystems behaviour if the external force actually influences only a single type of the components of the corresponding quasi-equilibria. In this case, variation of the mass (concentration) of these components can be easily measured. For example, the influence of magnetic fields on biological objects can involve only structures of certain type, with sizes exceeding some critical value. A good illustration is the destruction of aterosclerotic formations by the magnetic field and so on.

The law described by the semi-logarithmic equation (15) is actually known for a long time. It was formulated empirically and called the Weber-Fechner law [58]:

$$\ln Y = bZ + \ln Y_0 \ , \tag{16}$$

where Z is defined by some authors as the "sensation strength" (intensity of sound, taste, feelings, etc.); Y is introduced as the strength (energy, concentration) of the "irritator"; Y_0 is the threshold strength of the irritator; b is constant. Evidently, a nonstrict definition of the parameters of the Weber-Fechner law caused a lot of controversies. Other empirical dependencies were suggested as alternatives of this law, such as the Stievens exponential law [58], which is rather popular. However, this law is also not general and describes only particular cases. A more general law can be obtained in the framework of equilibrium thermodynamics (more specifically, macrothermodynamics). This law can explain the deviations from semi-logarithmic dependencies observed in the behaviour of biosystems at high intensities of the external irritation, when the systems cannot retain their stability and so on.

One can illustrate the general law (15), as well as the Weber-Fechner law, by the relation between the intensity of odour felt by human and the concentration of odoriferous substance in the air. In this case, the result of the external action is directly connected with the response of sensor systems, which can be considered as quasi-closed at small times. As an example, let us consider the relation between the intensity of odour, Z, and the logarithm of the concentration of ethylmercaptan in the air, c_s (Fig.12). The values of Z are given in a six-grade scale, and the concentration c_s is defined as the number of portions of the odoriferous substance per million portions of the air [59]. With an account for the mechanisms taking place in the receptors, one can conclude that Z is proportional to the logarithm of the substance concentration in the active zone of the receptors responsible for the influence of order. It turns out that the dependence presented in Fig.12 is indeed described by the law (15).

The Weber-Fechner law is also well satisfied for the relation between the felt intensity of sound, l, and its intensity, I :

$$\ln I = b'l + \ln I_0 , \qquad\qquad (17)$$

where I_0 is the sound intensity at the threshold of hearing, b' is constant [60]. The dependence (17) is functionally identical to the law (13) if one sets $K_i^j = I / I_0$, and $const=0$. The ratio I / I_0 is a dimensionless parameter, and I is connected with some concentration,

c_k, of the component corresponding to the current (instant) quasi-equilibrium. Evidently, for a certain range of the sound intensity the following linear relation is valid: $c_k = \gamma_k I$, where γ_k is the proportionality coefficient. The sound loudness I has the dimension of pressure and is an intensive factor (X_i), while the energy of the sound, $\Delta\Phi_\alpha$, is expressed in terms of the volume energy density. Hence, it follows that the extensivity factor (x_i) can be understood as the volume influenced by the sound energy. This volume should be a part of the active area in the hearing organ. Since the size of this zone can be assumed to be constant, the extensive parameter x_i is also constant; the only variable value is the intensive parameter, I. It follows from all these considerations that (13) and (17) can be considered as identical.

The Weber-Fechner law is well satisfied for the action of light on the organs of sight. That is why it was first obtained by astronomers during the study of the space bodies luminosity.

Equations analogous to (13) can be of great importance for pharmacology, therapy, anaesthesiology, sport physiology and some medical-physical-chemical disciplines while determining the optimal dosages of medicines or physiological actions [9, 61]. Approaching the maximal dosages or exercises should lead to deviations from the semi-logarithmic law since in this case the physiological systems of an organism cannot compensate the external influence. There are a lot of examples in pharmacology when relations are satisfied that can be considered as following from the general law (16). The corresponding examples can be found in monographs [62-64]. In accordance with the material presented above, we can hope that the appearance of papers [2-4] would initiate the birth of a new branch of science - pharmacothermodynamics. Of course, comparatively strict thermodynamic approach can be used in this area only under the condition that thermodynamics can be easily "separated" from kinetics. For instance, this is possible when a preparate is quickly injected into an organism, so that the variation of the medicine concentration can be neglected in the chosen time scale, whereas the action of the medicine manifests itself rather quickly. Generally, it should be noted that since biological systems are complicated and the models used sometimes are non-adequate, one should be extremely careful while applying to biology the thermodynamical law of least compulsion.

6. PROLONGATION OF LIFE AND THERMODYNAMICS

If the process of ageing is identified with the ontogenesis [25], then the degree of the ontogenesis accomplishment (extent of ontogenesis) is simultaneously the criterion of the ageing degree (of the biological age) of organisms. The concept of the equivalence between the extent of ontogenesis and ageing is in agreement with the thermodynamics of the evolution of organisms. Indeed, it is impossible to separate the "thermodynamics" of development, growth and ageing as parts of the process. Still, they are widely thought to be independent components of the ontogenesis.

Studying the ageing (ontogenesis) of the supramolecular structure of a biotissue, an organ, or any biosystem, it is reasonable to investigate how physical and chemical factors influence this process. These factors are usually assumed to include temperature, pressure, energetic content and composition of food, the effect of synthetic chemicals and natural compounds with physiological action, physical exercise, the spectrum and intensity of light, the effect of ionising radiation, physical fields, etc. If the aging (ontogenesis) takes place against the background of all these factors being constant (the factors should be regarded as parameters of the thermostat, i.e., of the habitat), then the ontogenesis of the bio-system is accomplished at a quite definite value of $\overline{\overline{G}}_j^{im}$ (1), which in some cases is close to its minimal value.

The "average life-span" of an organism is related to the species it belongs to and to certain averaged conditions of the habitat. Therefore, each species of living beings has a strictly "programmed" life-span corresponding to its typical habitat. However, the change of the habitat (or, in terms of thermodynamics, the change of the thermostat that takes place, for instance, for the changing the composition of food or for the cases of cancer or neoteny) makes the system tend to a new value of $\overline{\widetilde{G}}_j^{im}$ (2), which can be larger or smaller than $\overline{\overline{G}}_j^{im}$ (1). If some changes of the habitat occur at some moment of the organism's life, so that $\overline{\widetilde{G}}_j^{im}$ (2) $< \overline{\widetilde{G}}_j^{im}$ (1), then it is quite probable that the life-span will increase. Fluctuations of the parameters of the surrounding (Fig.13) can lead to the rejuvenation or aging of an organism within certain limits, which are called the adaptive zone. The transition from the curve 1^- to the curve 1^+ means, from the view point of thermodynamics, *rejuvenation*, while the transfer from the curve 1^+ to the curve 1^- means *aging*.

Thus, the rejuvenation of a certain organism, an organ, a functional system or some local zone of a biotissue, under the condition that the genetic characteristics are constant, is possible only due to the changes of the habitat conditions (parameters). Moreover, fluctuations of the habitat parameters such as atmospheric changes, variation of physical fields, etc., turn to cause the fluctuations of $\overline{\overline{G}}_j^{im}$. They "rejuvenate" or "age" the biotissues of an organism within the limits of the adaptive zone (zone of adaptive possibilities) and are manifestations of the thermodynamic "force" of the surrounding in the course of ontogenesis. The analogous scheme (Fig.13) can be applied to philogenesis and long periods of the biological evolution.

If the nutrition regime of an individual is changed, so that the amount of unsaturated fatty acids in the diet increases, then lipid-containing supramolecular structures (tissues) can be rejuvenated. The Gibbs function of the formation of such structures ($\Delta \overline{\overline{G}}_j^{im}$) will become less negative (transition from the curve 1^- to the curve 1^+). This statement following from the thermodynamic theory of ageing agrees with the evidence obtained in clinical practice while treating people for atherosclerosis with drugs like Linaetholum obtained from flax oil. The drugs contains a mixture of ethyl ethers of oleic, linoleic and linolenic acids. These fatty acids have low melting temperatures and positive values of $\Delta \overline{\tilde{G}}^{im}$ at 25°C (see Fig.10 and Table 2). When fatty acids and fats with high melting temperatures are forced out by analogous substances with low melting temperatures, the bio-tissues get rejuvenated. All this is in full accordance with the laws of thermodynamics! There are known other similar examples relating to the rejuvenation of fats, collagen, and other tissues. The observed facts give us hope that the appearing gerontological thermodynamics will help to develop new methods and synthesise new preparations that would slow down the process of ageing for human and other living beings.

One of the applications of our theory is connected with the problems of gerontology, dietology, and several other medical and biological sciences. It can be formulated in the form of the following principle:

Diets including «thermodynamic evolutionary young» animal and vegetable foods stimulate longevity and improve the quality of human life. The degree of «evolutionary» youth of a food product is determined by its chemical composition and supramolecular structure. The chemical composition and supramolecular structure of a product depend, in their turn, on its ontogenetic and philogenetic ages. An important quantitave measure of the «gerontological efficiency» of a food product is its the Gibbs function of supramolecular structure formation, which characterises the thermodynamic stability of its supramolecular structure.

Besides the gerontological effect of medical preparations on the aging of human biological tissues can be evaluated through testing which adds to the tool kit of evaluation of the quality of medicines.

Many medical preparations (vitamins, trace elements, hormone imitators and other biologically active compounds) help sustain the physiologically optimal stability of the supramolecular structures of the biological tissues of the organism. That optimal stability (estimated by the value of the Gibbs function of the formation of the structure) facilitates normal metabolism, slows down aging and improves the quality of life. Appropriate tests make it possible to identify the most effective preparations and recommend their doses from the point of view of gerontology. A geronotological screening of a preparation can rely on the known data about the chemical structure of the medicine, composition and the structure of the supramolecular structures of biological tissues where the preparation needs to be localized (lipid structures, membranes, serum, collagen tissue, etc.) and on simple physical and chemical tests such as solubility in standard systems, etc.

To conclude this Section, we shall make the following remarks.

Above, in Fig.4, we have shown the scheme representing the averaged tendencies in the variation of $\Delta\widetilde{\overline{G}}^{ch}$ ($\widetilde{\overline{G}}^{ch}$) and $\Delta\widetilde{\overline{G}}^{im}$ ($\widetilde{\overline{G}}^{im}$) of the biotissue in the course of onthogenesis. Similar averaged tendencies are shown in Fig.7 not only for ontogenesis but also for philogenesis. In both cases, the values of $\Delta\overline{\overline{G}}^{ch}$ and $\Delta\overline{\overline{G}}^{im}$ were averaged over the volume of the biotissue, which is a heterogeneous system. Besides, the scheme in Fig.4 show oscillations in time of $\Delta\widetilde{\overline{G}}^{ch}$ and $\Delta\widetilde{\overline{G}}^{im}$

connected with the possible variations of the environment parameters. The scheme depicted in Fig.13 also takes into account that the values $\Delta\overline{\widetilde{G}}^{ch}$ and $\Delta\overline{\widetilde{G}}^{im}$ averaged over the biotissue volume can oscillate due to the variations of the environment parameters.

Studying the biotissue on the "microscopic level", one can see that since the system is heterogeneous, the microvolumes, which still contain a sufficiently large number of particles, are characterized by different values of $\Delta\overline{\widetilde{G}}^{ch}$ and $\Delta\overline{\widetilde{G}}^{im}$.

Let us select a plane cut of a biotissue with thickness equal to the size of microvolumes, which are of cubic shape. If every microvolume of this cut is characterized by the value $\Delta\overline{\widetilde{G}}^{im}$, then one can easily plot the distribution of $\Delta\overline{\widetilde{G}}^{im}$ over the cut in the form of a topographic map. This map resembles a mountain relief. Such reliefs of $\Delta\overline{\widetilde{G}}^{im}$ are presented in Fig.14 for four chosen moments of the biotissue growth. The fact that the areas with hills (which characterize different stages of ontogenesis) grow in size indicates that any volume of the biotissue selected initially would grow in size (inflate) as the organism grows.

Similar reliefs can be also built for $\Delta\overline{\widetilde{G}}^{ch}$ or ΔH_{comb}^{ch} (the heats of combustion).

Fluctuations of the parameters of the environment (thermostat), such as temperature, pressure, chemical composition of food, etc., lead to various changes in the reliefs of and $\Delta\overline{\widetilde{G}}^{im}(t)$ and $\Delta\overline{\widetilde{G}}^{ch}(t)$ (Fig.14). However, these changes are limited by the possibility of the adaptive zones. A biotissue (an organism) can adapt to the variations in the parameters of the thermostat only if these variations lie within the limits of the adaptive zones.

Considerations presented above are valid for any living system: an organell, a cell, an organ, an organism, a population, etc. One can create topographic maps fixing the values of $\Delta\overline{\widetilde{G}}^{im}$ and $\Delta\overline{\widetilde{G}}^{ch}$ at any moment of the evolution of a biosystem. Further, it is easy to make averaging over fixed volume and time, and build dependence like those presented in Figs. 4,7.

It is also possible to build three-dimensional schemes that could show variations of $\Delta\overline{\widetilde{G}}^{im}$ and $\Delta\overline{\widetilde{G}}^{ch}$ in the microvolumes of biotissues. One of such schemes is presented in Fig.15.

Thus, the values $\Delta\overline{\widetilde{G}}^{im}$ and $\Delta\overline{\widetilde{G}}^{ch}$ (or the components of the specific Gibbs function of the formation of complicated systems) vary in the volume of a biosystem and oscillate in time.

If an organism is considered, then the variation of $\Delta\overline{\widetilde{G}}^{im}$ and $\Delta\overline{\widetilde{G}}^{ch}$ in the volume is observed at any time moment on the level of microvolumes (supramolecular structures forming organells, biotissue cells, the organism itself) and also macrovolumes of biological objects relating to different hierarchies.

Time oscillations of $\Delta\overline{\widetilde{G}}^{im}$ and $\Delta\overline{\widetilde{G}}^{ch}$ in the microvolumes and macrovolumes of the biomass can be connected, as we have already mentioned, to the changes in the parameters of the environment surrounding the biosystem under consideration (a fragment of supramolecular structure of a cell, the cell itself, etc.). These oscillations take place against the background of constant thermodynamic tendencies, which are shown in Figs. 4,7 and 13.

Thermodynamics admits averaging to be performed on the level of macrovolumes of biosystems and at sufficiently large times. It turns out that for biological objects of all hierarchies (a macrovolume containing microvolumes of supramolecular structures, an organell, a cell, an organ, an organism, a population, etc.) the average values $\Delta\overline{\overline{G}}^{im}$ and $\Delta\overline{\overline{G}}^{ch}$ should tend to their extreme values.

To be sure, biosystems of all hierarchic levels can be characterised by their average (with respect to volume or mass) chemical composition. As it was several times mentioned in this work, for a long time it has been a customary way in biology to compare the averaged chemical compositions of cells, tissues, organs, organisms, populations and other biological objects while discussing the problems of living systems evolution.

7. WHY ARE NUCLEIC ACIDS THE CARRIERS OF GENETIC INFORMATION ?

It is commonly accepted that nucleic acids and above all DNA are the carriers of genetic information. The birth of every living object, i.e., a cell starts with a nucleic acid.

Why are these natural macromolecules uniquely qualified to retain and carry genetic information ?

If the author's theory about the thermodynamic nature of evolution from ontogenesis to phylogenesis is to be accepted than we must accept that the part of nucleic acids and in particular DNA in the origination and evolution of life on Earth was preordained . Nucleic acids could fulfill their mission on Earth or another planet with liquid water and a composition of the atmosphere similar to that of a juvenile Earth. One must also bear in mind that DNA is framed with proteins and other chemical components of various origins. That framing forms appropriate supramolecular structures around the double spiral of DNA and these structures possess a measure of thermodynamic stability. The stability of the DNA spiral is therefore determined not only by the stability of the AT and GC pairs but also by the nature of the molecular and ionic frame around the double spiral. Therefore it makes sense to discuss from the thermodynamic perspective only the principal role played by DNA as the carrier of genetic information whereas the properties and functions of DNA are also determined to a certain extent by the chemical and supramolecular structures framing the double spiral.

Let us attempt briefly to answer the question posed in the heading.

The structure of the double spiral as compared to other supramolecular structures is distinct among other things by its greater thermodynamic stability which can be estimated by quantity ΔG^{im}. The melting point θ_{δ} of the double spiral (and chromatin) is relatively high. It is considerably higher than that of lipids and many proteins. Clearly, the higher value of $\left| -\Delta G^{im} \right|$ assures high accuracy of self-replication, prolonged stability and relatively high endurance with regard to mutations. Of all biological structures DNA and the framing around it are the most conservative.

At the same time the chemical stability of purine bases (A, G), pyrimidine bases (T, C), adenosine phosphoric acids (and similar compounds) is low and their chemical energy content is relatively high. The low chemical stability of the monomeric links of nucleic acids determines the crucial features of the metabolism and facilitates the rapid rate of their exchange in an organism.

It is therefore the thermodynamic conservatism of DNA deriving from the relatively slow change of its structure in a constantly reproducing biological mass that determined the role of that biopolymer as the carrier of genetic information.

There are grounds to believe that the unique role of nucleic acids correlates macrothermodynamically with the " principle of stability of chemical substance " relating the chemical and supramolecular stability of condensed substances (see Appendix 2).

8. ON THE SEEMING CONTRADICTIONS BETWEEN THE MODEL AND THE REALITY

It is worth paying attention to the questions that usually arise when researchers first get acquainted with the thermodynamic model of biological evolution.

For instance, an interesting question is whether the growth of a cancer tumour contradicts to the model. This question is reasonable because the development of cancer is accompanied by the rejuvenation of cells with increasing water content.

One of the possible answers is quite simple.

Rejuvenation of cancer cells is connected with their accelerated regeneration at early stages (in comparison with normal cells). A cancer tissue gets enriched with young cells where the content of water, naturally, is high. Such a tissue simply contains a larger number of young cells. Ontogenesis seems to slow down. However, each young cell gets older, though this process is comparatively slow. Thus, the cancer tissue develops in accordance with thermodynamics. As to the factors causing the appearance of a tumour (cancer cells), they have not only thermodynamic nature but can be connected with heredity and kinetics (mechanisms of processes). In the present work, the mechanisms of this phenomenon are almost not discussed (see p. 51).

Another question: why ancient species of living beings undergo slow evolution during philogenesis, while the evolution of new species is faster? This question usually arises while considering dependencies as shown in Fig.9. The question relates, first of all, to the mechanisms of the phenomenon and has no direct connection with thermodynamics. However, it should be pointed out that the habitats (thermostats) of ancient species, apparently, change rather slowly. New species live in habitats that vary faster, and hence they have to adapt rapidly to the new conditions. This should accelerate their evolution. Thus, in this case thermodynamics "manifests itself" in the effects accompanying the change of the thermostats (habitats) surrounding living beings.

Sometimes the author is reproached for a too large generality of the model and also for a lack of experimental facts confirming the tendencies presented in Figs.4 and 7.

Accepting the viewpoint that thermodynamics is the motive force of the evolution, we come to the conclusion that the model should be as general as the thermodynamic method is.

As to the experimental facts confirming the model, one can state that the number of obtained dependencies similar to those shown in Figs.1-3 is more than enough for an unambiguous conclusion about the tendencies in the directed variation of the chemical composition and hence, $\Delta\overline{\widetilde{G}}^{im}$ and $\Delta\overline{\widetilde{G}}^{ch}$, in the course of ontogenesis, philogenesis, and the evolution in general. As it has been mentioned above (p.51), calculations of $\Delta\overline{\widetilde{G}}^{im}$ and $\Delta\overline{\widetilde{G}}^{ch}$ can be easily carried out using the known experimental data [33, 67]. Numerous calculations made by the author are in full agreement with the model presented in the work (Fig.4).

There are some problems that require further studies. For instance, there is still no thermodynamic data describing the enrichment of biological structures with water at early stages of their life cycle. These are the processes of seeds swelling during their germination, maturation of an ovum and so on. In these cases, the environment of a system changes dramatically, which is equivalent to the "transfer" of systems into new thermostats (see Fig.7). For instance, a wheat seed put into wet soil would first, according to the laws of phase equilibria, get enriched by water. Next, the seed spouts, and the plant develops according to the scheme in Fig.4 (see also the stages of ontogenesis, Fig.7).

The problem of selecting a standard state characterized by values of $\Delta\overline{\widetilde{G}}^{ch}$ in comparing the energy capacity of open systems is a well known one. The difficulties in determining the energy content can be resolved to a degree if the energy capacity is estimated by means of the value of ΔI^0_{comb} as is done in dietetics. On the whole, an accurate comparison of the energy capacity and thermodynamic stability of systems of inherently different chemical composition is impeded because it is impossible to determine the absolute values of the functions G, F, H and some others.

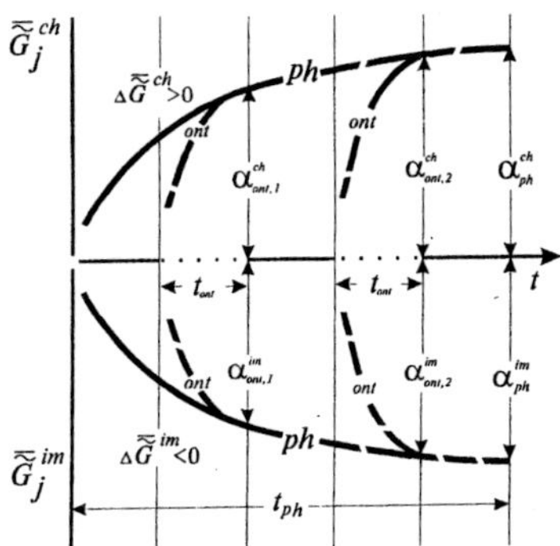

Fig. 7. Schematic plot of the variation of the specific values constituting the Gibbs function of the biosystem j, which is the biotissue of an animal, in the course of ontogenesis and philogenesis. (The extent of ontogenesis or philogenesis is difined by the content of water in the biotissue). This scheme can be also applied to durable periods of the biological evolution.

$\overline{\overline{G}}_j^{\,ch}$ and $\overline{\overline{G}}_j^{\,im}$ are the specific values of the chemical (ch) and supramolecular (im) components relating to the j-th system; the specific Gibbs function of the matter is $\overline{\overline{G}}_j = \overline{\overline{G}}_j^{\,ch} + \overline{\overline{G}}_j^{\,im}$; t_{ont}, t_{ph} are the time scales of ontogenesis and philogenesis, respectively; the scales $\overline{\overline{G}}_j^{\,ch}$ and $\overline{\overline{G}}_j^{\,im}$ are dimensionless ($\overline{\overline{G}}_j^{\,ch} >> \overline{\overline{G}}_j^{\,im}$); the functions $\overline{\overline{G}}_{j,ont}^{\,ch} = f_1(t_{ont})$, $\overline{\overline{G}}_{j,ph}^{\,ch} = f_1(t_{ph})$, $\overline{\overline{G}}_{j,ont}^{\,im} = f_2(t_{ont})$, and $\overline{\overline{G}}_{j,ph}^{\,im} = f_2(t_{ph})$ are arbitrary. The values of α with subscripts and superscripts characterize the components of the specific Gibbs function of structure formation (the specific Gibbs free energy of chemical and supramolecular structure formation) for the system at some fixed moment, t. The values α vary in the course of the evolution (ontogenesis and philogenesis) due to the changes in the chemical composition of organisms or species. The values of the Gibbs function are differences of the Gibbs functions for two different states, one of them being taken as a standard, as usual. According to this, the values $\overline{\overline{G}}_j^{\,ch}$ and $\overline{\overline{G}}_j^{\,im}$ can be plotted as ordinates instead of $\Delta\overline{\overline{G}}_j^{\,ch}$ and $\Delta\overline{\overline{G}}_j^{\,im}$ (see Fig.4).

Besides, it is more convenient in practice to use values of $\Delta\overline{\overline{H}}_{comb}^{\,ch}$ instead of $\Delta\overline{\overline{G}}_j^{\,ch}$.

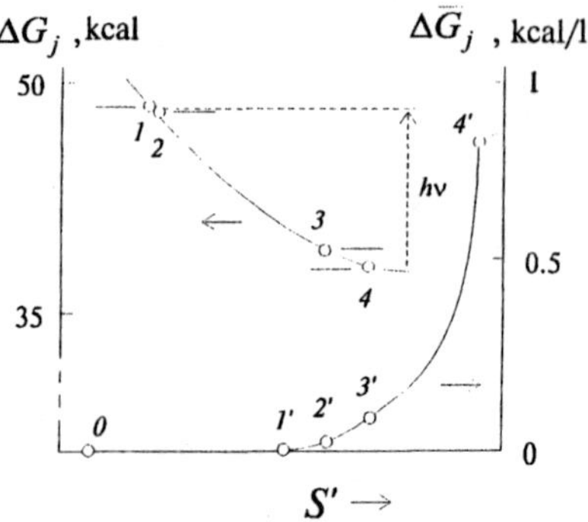

Fig.8. The Gibbs function formation, ΔG_j (the Gibbs function variation) for a model closed polyhierarchic system j and the volume density of the Gibbs function of the formation, $\Delta \overline{G}_j$ for quasi-closed subsystems of the system j. Both values are plotted against the evolution coordinate S' ($n = 0, 1, 2, ...$).

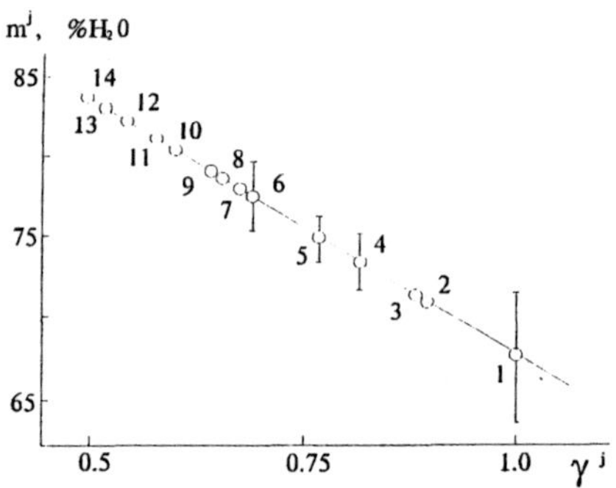

Fig.9. The amount of water contained in the brain as a function of its relative development degree for the j-th animal (γ^{j}). It is supposed that γ^{j} depends linearly on the concentration of water in the brain.

It is assumed for human $\gamma^{m}=1$ ($j \equiv m$ - man), for frog $\gamma^{f} = 0.5$ ($j \equiv f$ - *frog*). The average data are presented for adult animals: 1- human; 2 - monkey; 3 - horse; 4 - dog; 5 - cat; 6 - rabbit; 7 - duck; 8 - mouse; 9 - guinea pig; 10 - rat; 11 - shark, carp , seagull; 12 - perch; 13 - turtle; 14 - frog.

Table 1[*]

Specific standard Gibbs functions of the formation of natural compounds from elements at 25°C ($\Delta \overline{G}_{f_{298}}{}^0$)

Compound	Formula	State	$\Delta \overline{G}_{f_{298}}{}^0$, ccal/g
Adenine	C5H5N5	S	+0.53
Guanine	C5H5N5O	S	+0.075
1,2,3-Propanetriyl trimyristate	C45H86O6	L	-0.19
Palmitic acid	C16H32O2	S	-0.29
Phenylalanine	C9H11NO2	S	-0.31
Arginine	C6H14N4O2	S	-0.33
Tyrosine	C9H11NO3	S	-0.51
Leucilglycine	C8H16N2O3	S	-0.60
Leucine	C6H13NO2	S	-0.63
Cystine	C6H12N2O4S2	S	-0.69
Alpha-lactose monohydrate	C12H24O12	S	-1.16
Glycine	C2H5NO2	S	-1.18
Glutamic acid	C5H9NO4	S	-1.18
Alpha-galactose	C6H12O6	S	-1.22
Glucose	C6H12O6	S	-1.22
Glycerine	C3H8O3	L	-1.24
Water	H2O	L	-3.147

[*] The table contains examples of chemical compounds used by Nature for building molecular (*ch* or *m*) and supramolecular (*im*) structures of living beings. $\Delta \overline{G}_{f_{298}}{}^0 = \Delta \overline{G}_{f_{298}}^{ch}{}^0 + \Delta \overline{G}_{f_{298}}^{im}{}^0$. The values of $\Delta \overline{G}_{f_{298}}{}^0$ are calculated according to [27].

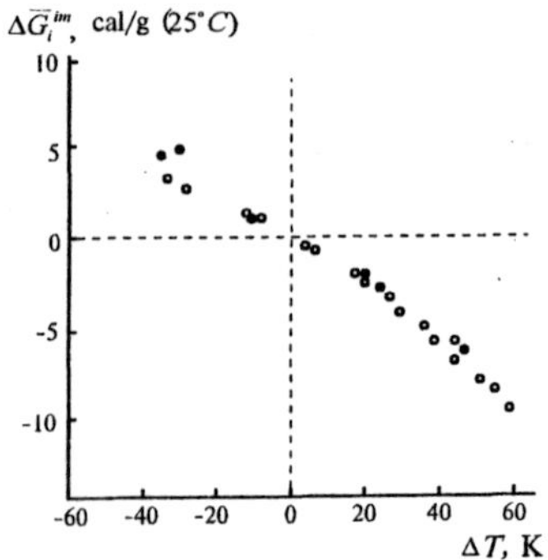

Fig.10. The specific Gibbs function $\Delta \overline{G}_i^{im}$ of non-equilibrium phase transition "supercooled liquid — solid" as a function of ΔT ($\Delta T = T_{m_i} - 298.2K$) at 298K for a series of fatty acids; $\Delta \overline{G}_i^{im}$ and T_{m_i} are the specific Gibbs function of crystallization (condensation) and the melting temperature of the i-th compound, respectively. The value of $\Delta \overline{G}_i^{im}$ is calculated per unit mass. The correlation does not vanish when $\Delta \overline{G}_i^{im}$ is calculated per unit volume. Empty circles (∘) relate to the saturated fatty acids, filled circles (•) to the unsaturated fatty acid.

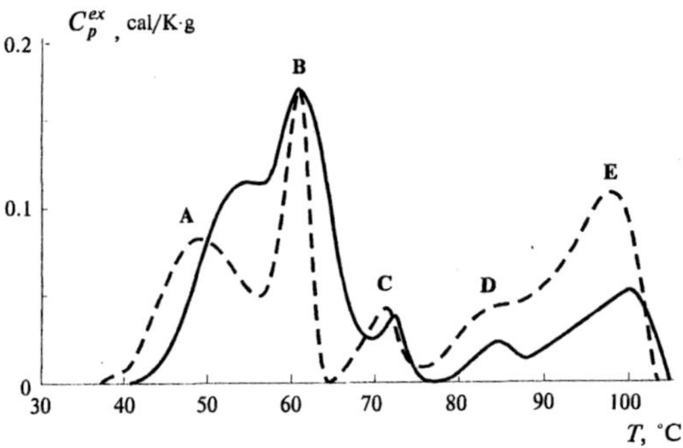

Fig.11. The DSC profiles ($C_p^{ex} = f(T)$) of "HeLa" cells (solid curve) and of "V79' cells (dashed curve) [42].

Table 2

Several fatty acids whose residuals are contained in natural fats[*]

Acid	Mol. mass	T_m, °C	ΔH_m, cal/g	ΔT, K	$\Delta \overline{G}^{im}$, cal/g	$\Delta \overline{G}^{im}$, cal/mol
Saturated						
n-Lauric (25 °C)	200.3	43.2	43.7	18.2	-2.52	-503
Myristic (25 °C)	228.4	54.0	47.5	29.0	-4.22	-964
Palmitic (25 °C)	256.4	61.8	50.7	36.8	-5.57	-1428
Palmitic (5 °C)				56.8	-8.60	-2204
Stearic (25 °C)	284.5	68.8	52.6	43.8	-6.74	-1917
Unsaturated						
Oleic (9c) (25 °C)	282.5					
(α)		13.4	25.5	-11.6	+1.03[**]	+292[**]
(β)		16		-8.7	+0.767[**]	+217[**]
Oleic (9c) (5 °C)						
(α)		13.4	25.5	8.4	-0.747	-211
(β)		16		11.3	-0.996	-281
Elaidic (9t) (25 °C)	282.5	44.4	33.0	19.4	-2.02	-569
Elaidic (9t) (5 °C)				39.4	-4.09	-1157
Linoleic (9c, 12c) (25 °C)	280.5	-5.1	43.9	-30.1	+4.93[**]	+1383[**]
Linoleic (9c, 12c) (5° C)				-0.1	+0.016[**]	+4.59[**]

[*] Values of T_m and ΔH_m are taken from reference literature [16]. The estimations for $\Delta \overline{G}^{im}$ are done using Eq. (7) for the case of crystallisation (hardening) of substances from over-cooled state at 25 and 5°C.

[**] The values of $\Delta \overline{G}^{im}$ are positive since the standard temperature 25°C (5°C) is higher than T_m. Under this condition, crystallisation of matter is impossible.

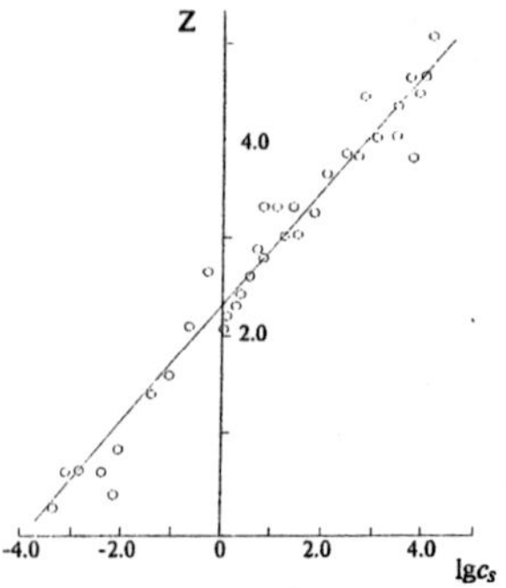

Fig. 12. Relation between the odour intensity (Z) and the logarithm of the concentration of ethylmercaptane in the air [59]. c_s is the number of portions of the substance per million portions of air.

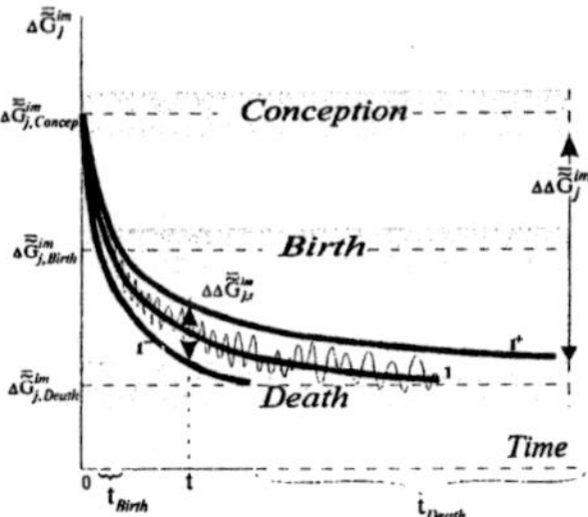

Fig. 13. Schematic variation of the Gibbs function corresponding to the formation of the aggregated phase of supramolecular structures for the biotissue of an organ of the j-th organism ($\Delta \overline{\overline{G}}_j^{im}$) in the course of ontogenesis.

The curve 1 describes the organism j living under given (standard) conditions of the habitat of (human) population. The dependence is determined by the genetic characteristics of the organism (genotype) j and by the given conditions of its habitat, i.e., the averaged parameters of the thermostat. The curves 1^+ and 1^- correspond to the individual j under the assumption that it lives under extremal conditions, which are different from the given ones. The area between the curves 1^+ and 1^- corresponds to the adaptive zone of $\Delta \overline{\overline{G}}_j^{im}$ variations due to the fluctuations of the habitat parameters. Within the limits of this zone, rejuvenation or accelerated aging is possible for the object (for instance, the biotissue or the whole organism).

The upper dashed line *(Conception)* relates to the conception of an individual (genotype) j, the value $\Delta \overline{\overline{G}}_{j,Concep}^{im}$ being predetermined genetically, corresponding to the habitat of the population.

The middle dashed line *(Birth)* relates to the birth of the animal (the man) j.

The lower dashed line *(Death)* corresponds to the death of the organism j; the death time can vary within the adaptive zone. It is determined by the changes in the habitat conditions.

The value $\Delta\Delta \overline{\overline{G}}_j^{im}$ is the variation of the specific Gibbs function for the biological object j during ontogenesis (as a result of aging) from conception to death. For example, for the aging of the collagen tissue of animals, $\Delta\Delta \overline{\overline{G}}_j^{im} \cong -0,3 \, cal/g$. The variation $\Delta\Delta \overline{\overline{G}}_{j,t}^{im}$ characterizes the adaptive zone width at the moment t of the organism's life.

The values $\Delta\Delta\overline{\widetilde{G}}_j^{\,im}$ (calculated with respect to some standard level, $\Delta\overline{\widetilde{G}}_{j,st}^{\,im}$) for the lines *"Birth"* and *"Death"* can differ from the average values $\Delta\Delta\overline{\widetilde{G}}_{j,st}^{\,im}$, which characterize an "average" individual living in "average" standard habitat (*st*). These differences can be described by the variations $\Delta\Delta\Delta\overline{\widetilde{G}}_{j,st}^{\,im}$ *(Birth)* and $\Delta\Delta\Delta\overline{\widetilde{G}}_{j,st}^{\,im}$ *(Death)*.

The value $\Delta\Delta\Delta\overline{\widetilde{G}}_{j,st}^{\,im}$ can have different sign. It is a quantitative characteristics of the aging extent for the given object *j* at the moment of his life (*t*) or at the death moment. The shaded areas surrounding the dashed lines *"Conception"*, *"Birth"*, and *"Death"* denote the zones where the values $\Delta\Delta\overline{\widetilde{G}}_{j,Concep}^{\,im}$, $\Delta\Delta\overline{\widetilde{G}}_{j,Birth}^{\,im}$, and $\Delta\Delta\overline{\widetilde{G}}_{j,Death}^{\,im}$ fluctuate. (The sizes of the zones are comparable with $\Delta\Delta\Delta\overline{\widetilde{G}}_{j,Concep}^{\,im}$, $\Delta\Delta\Delta\overline{\widetilde{G}}_{j,Birth}^{\,im}$ and $\Delta\Delta\Delta\overline{\widetilde{G}}_{j,Death}^{\,im}$).

The saw-tooth line plotted against the curve 1 symbolical emphasizes that the fluctuation of the parameters of the surrounding, such as temperature, pressure, nutrition schedule, physical fields, the change of day and night, the change of seasons, etc., lead to the variation of $\Delta\overline{\widetilde{G}}_j^{\,im}$. The organism adapts to these variations only within the limits of the adaptive zone.

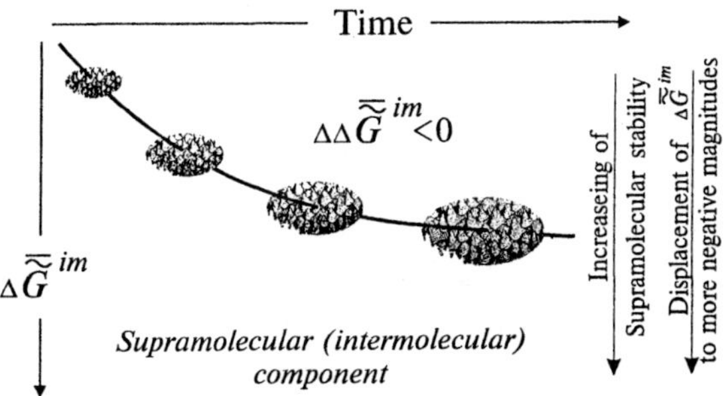

Fig.14. Topographic maps of $\Delta\overline{\widetilde{G}}^{\,im}$ for cuts of a biotissue in the process of ontogenesis. The sizes of areas with "hills" shown for fixed time moments characterize how the initial volume of biotissue "inflates". In the course of ontogenesis, the value $\Delta\overline{\widetilde{G}}^{\,im}$ becomes more negative ($\Delta\Delta\overline{\widetilde{G}}^{\,im} < 0$), while the stability of supramolecular structures increases.

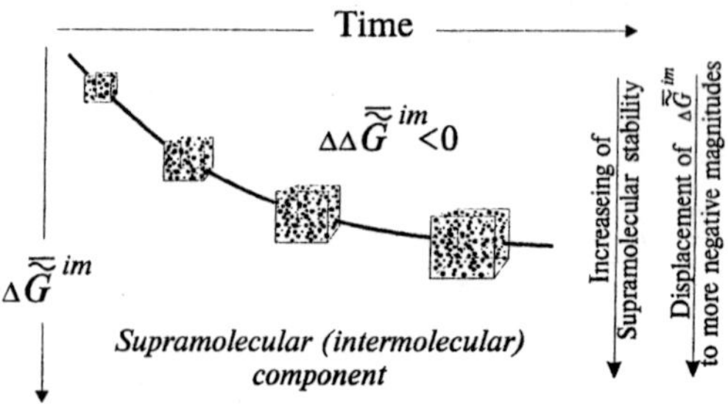

Fig.15. The scheme showing oscillations of $\Delta\overline{\widetilde{G}}^{im}$ in the microvolumes of a biotissue (cells, organells, mozaic structuras of membranes, etc.). The increase of the cube size is a manifestation of the fact that an initial microvolume of biotissue selected, for instance, at the moment of birth, "inflates" in the course of ontogenesis. The number of dark spots in each cube corresponds to the number of selected microvolumes in the biomass. The size of the spots is proportional to the values of $\Delta\overline{\widetilde{G}}^{im}$ in the corresponding microvolumes. In the course of ontogenesis, the value of $\Delta\overline{\widetilde{G}}^{im}$ of the biomass becomes more negative ($\Delta\Delta\overline{\widetilde{G}}^{im}< 0$), while the stability of supramolecular structures increases.

REFERENCES

1. *Gibbs J.W.* (1928). *The Collected Works of J. Willard Gibbs*. Thermodynamics. V.1. New York: Longmans, Green and Co., 55-349 p.

2. *Gladyshev G.P.* (1977). *Thermodynamics of Biological Evolution*, Preprint, Chernogolovka, Inst. Chem. Phys., 46 p.

3. *Gladyshev G.P.* (1978). On the Thermodynamics of Biological Evolution *J.Theoret.Biol.*, v. 75, p.425.

4. *Gladyshev G.P., Ershov Yu.A., Loshchilov V.I.* (1980). Thermodynamic Principles of the Biological systems. *J.Theor.Biol.*, v. 83, p.17.

5. *Gladyshev G.P.* Classical Thermodynamics, Tandemism and Biological Evolution. *J.Theor.Biol.*, v.94, p. 225.

6. *Gladyshev G.P., Ershov Yu.A.* (1982). Principles of the Thermodynamics of Biological Systems. *J. Theor. Biol.*, v. 94, p. 301.

7. *Gladyshev G.P.* (1983). *Hierarchic Thermodynamics of Complex Open Systems,* Preprint, Ufa: Bashk. Fil. USSR Academy of Sciences.

8. *Gladyshev G.P.* (1987). On the Macrokinetics and Thermodynamics of Natural Hierarchic Processes. *J. Phys. Chem.,* v.61, N 9, p. 2289.

9. *Gladyshev G.P.* (1988). *Termodinamika I makrokinetika prirodnykh ierarkhicheskih processov* (Thermodynamics and Macrokinetics of Natural Hierarchic Processes). Moscow: Nauka, 287 p.

10. *Gladyshev G.P.* (1993). Macrothermodynamics of Biological Systems and Evolution. *J. Biol. Syst.,* v.1, N 2, p.115.

11. *Gladyshev G.P.* (1992). *Macrothermodynamics of Biological Systems and Evolution.* Preprint. M: Inst. Chemical Phys. Russian Acad. of Sciences, Institute of Ecological and Biolophysical Chemistry, Academy of Creative Endeavours.

12. *Gladyshev G.P.* (1994). The Motive Force of Biological Evolution. *Vestnik RAN,* v.64, N 3, p. 221.

13. *Gladyshev G.P.* (1994). Thermodynamics and Biological Evolution. *J. Biol. Phys.,* v. 20, p.213.

14. *Vakulichev L.V., Gladyshev G.P.* (1989). Verification of the Macrothermodynamic Model of Biological Evolution. *J. Phys. Chem.,* v.63, N 10, p.2751.

15. *Gladyshev G.P., Gladyshev D.P.* (1994). On the model of biological systems evolution. *Izv. Acad. Nauk. Ser. Biol.,* N 1, p. 14.

16. *Gladyshev G.P.* (1995). Thermodynamic Trends of Biological Evolution. *Biology Bulletin,* v.22, N 1, p.1. ISSN 1062-3590; N 1.

17._Denbigh K.G._ (1989). The Many Faces of
Irreversibility _Brit. Jour. Phil. Sci._, v.40, p.501.

18._Denbigh K.G._ (1989). Note on Entropy, Disorder and
Disorganisation _Brit. Jour. Phil. Sci._, v.40, p.323.

19._Gladyshev G.P., Popov V.A._ (1978). _Radikalische
Polymerisation bis zu hoken Umsatzen._ B.: Acad.-
Verlag., 228 S.

20._Gladyshev G.P._ (1995). Termodinamika Ierarkhicheskah
Sistem (Thermodynamics of the Hierarchic Systems).
In: _Khimicheskaya Enciklopedia_ (Chemical
Encyclopedia). M.: v.4, p.535.

21._Gladyshev G.P._ (1996). Thermodynamic Trends of
Biological Evolution. Model and Reality. _Biology
Bulletin_ ISSN 1062-3590; N 4.

22._Gladyshev G.P._ (1997). _Thermodynamics of Evolution
of Living Beings. AAAS Annual Meeting and Innovation
Exposition_ (AMSIE`97), Seattle, February 13-18.

23._Gladyshev G.P._ (1994). _Chemical Evolution: Structure
and Model of the First Cell. Conference held in
Trieste (Italy), 29 August - 2 September 1994._ (Ed.
C.Ponnamperuma and G.Chela-Flores), p. 213.

24._Gladyshev G.P._ (1995). _Trieste Conference on
Chemical Evolution, IV: Physics of the Origin and
Evolution of Life. Eighth Session. Biophysical
Aspects._ Miramare, Trieste (Italy), 4-8 September
1995.

25._Gladyshev G.P._ (1995). _Mini-International Symposium
on Frontiers of Life Sciences._ October 15-17,
Tsinghua University. Beijing, P.R.China.

26._Prigogine I._ (1980). _From Being to Becoming: Time
and Complexity and the Physical Sciences._ San
Francisco: W.H. Freeman & Co.

27._Stull D.R., Westrum E.F. Jr., Sinke G.C._ (1969). _The
Chemical Thermodynamics of Organic Compounds._ New

York, London, Sydney, Toronto: John Wiley & Sons Inc.

28. *Gladyshev G.P., Kitaeva D.Kh.* (1995). On the Thermodynamic Tendency of Evolution Processes. *Izv. RAN. Ser. Biol.*, N 6, p. 645.

29. *Gladyshev G.P., Gladyshev D.P.* (1996) On the Physico-Chemical Theory of Biological Evolution. *J. Biol. Syst.*, v.4, N 2, 239.

30. *Gladyshev G.P., Kitaeva D.Kh., Ovcharenko E.N.* (1996). Why Does the Chemical Composition of Living Things Adapt to the Environment. *J. Biol. Syst.*, v.4, p. 555.

31. *Kitaeva D.Kh, Gladyshev G.P.* (1997). On the Calculation of the Gibbs Function Variation at Non-equilibrium Phase Transitions. *Izv. RAN. Ser. Biol.*, (in press).

32. *Mazariegos Manolo, Wang Zi-mian et.al.* (1994). Differences Between Young and Old Females in the Five Levels of Body Composition and Their Relevance to the Two-Compartment Chemical Model. *J. Gerontology: Medical Sciences*, v. 49, N 5, p.201.

33. *Widdowson E.M.* (1968). In: Body Composition in Animals and Man. *Proc. Symp. Held May 4-6 1967.* Univ. of Missouri. Columbia Publ. 1598. Wash. (D.C.): Nat. Acad. Sci., p.72.

34. *Aleksandrov V.Ja.* (1975). *Kletki, makromolekuly i temperatura* (Cells, Macro-molecules, and Temperature). Leningrad: Nauka, 330 p.

35. *Schuster P.* (1993). RNA Based Evolutionary Optimization. *Origins of Life and Evolution of the Biosphere*, v. 23, p. 373.

36. *Tsereteli G.I., Belopol'skaya T.V.* (1994). New Data on Thermal Denaturation of fibryl-collagen. *Biofizika*, v.39, N 5, p. 793.

37.*Nikitin V.N., Persky E.E., Utevskaya L.A.* (1977).
*Age and Evolution Biochemistry of Collagen
Structures.* Kiev: Nauk. Dumka, 280 p.

38.*Tanford Ch.* (1994). The Hydrophobic Effect and the
Organization of Living Matter. In: *Origins of Life.
The Central Concepts.* Ed. Deamer D.W., Fleischaker
G.R. Boston-London: Jones and Bartlett Publ. Inc.,
p.233.

39.*Fox S.W. and Dose K.* (1977). *Molecular Evolution and
the Origin of Life.* Revised Ed. New York: Marcel
Dekker.

40.*Lepock James R., Kruuv Jack* (1992). Mechanisms of
Thermal Cytotoxicity. *Hyperthermic Oncology,* v.2,
p.9.

41.*Lepock James R., Frey Harold E., Ritchie Kenneth P.*
(1993). Protein Denaturation in Intact Hepatocytes
and Osolated Cellular Organelles During Heat Shock.
J. Cell Biol., v. 122, N 6, p.1267.

42.*Lepock James R., Frey Harold E., Senisterra
Guillermo A., Heynen Miriam L.P.* (1995). Mechanisms
of Thermal Damage. *Radiation Research* 1895 - 1995,
v.2.

43.*Ritchie K.P., Keller B.M., Syed K.M., Lepock J.R.*
(1995). Hyperthermia (Heat-Shock)-induced Protein
Denaturation in Liver, Muscle and Lens Tissue as
Determined by Differential Scanning Calorimetry. J.
Hyperthermia, v. 10, N 5, p.605.

44.*Pfeil W., Privalov P.* (1979). In: Jones M.N. (Ed.)
Biochemical Thermodynamics. Amsterdam, Oxford, New
York: Elsevier Scientific Publ. Co.

45.*Andronikashvili E.L., Monaselidze D.R.,
Chanchalashvili Z.I., Majachaladze G.V.* (1983).
Cooperative Thermal Transitions in Normal and Tumour
Cells. *Biofizika,* v. 28, N 3,. p. 528.

46. *Esipova N.G., Lazarev Yu.A., Tiktopulo E.I., Aizenhaber F. (1993). Experimental Evidence for the New Understanding of the Role of Amino-acids in the Thermodynamics of Collagen-Type Proteins. Biofizika,* v. 38, N 1,. p. 203.

47. *Katz J., Crespi H. (1970). Isotope Effects in Biological System. In: Isotope Effects in Chemical Reactions.* (Ed. Collins). New York: Van Nostrand, p. 286.

48. *Hubnen G., Jung K., Winkler E. (1970). Die Rolle Des Wassersin Biologischen Systemer.* Bereia: Academia Verlag.

49. *Gladyshev G.P.* (1996). Thermodynamics and the Evolution of Biological Structures. *Biofizika* (in press).

50. *Mrevlishvili G.M., Razmadze G.Z., Metreveli N.O., Kakabadze G.R.* (1995). Physics of the Interaction of DNA and Water with an Account for Two Types of Hydrogen Bonds in Water. *Biofizika,* v.40, N 2, p. 293.

51. *Bazarov I.P.* (1983). *Termodinamika.* (Thermodynamics). 3rd Ed. Moscow: Vysshaya shkola, 344 p.

52. *Sychev V.V.* (1981). *Differentsial'nye uravneniya termodinamiki* (Differential equations of thermodynamics). M.: Nauka, 195 p.

53. *Kandel E.R.* (1976). *Cellular Basis of Behavior.* San Francisco: W.H. Freeman and Co.

54. Novoseltsev V.N. (1978). *Teoriya upravleniya I biosystemy.* (Theory of Control and Biosystems.) M.: Nauka, 319 p.

55. Rakich L. (1984). *Regularnye sistemy povedeniya. (Regular systems of behaviour.)* M.:Mir, 134 p.

56. Svirezhev Yu.M., Logofet D.O. (1978). *Ustoichivost' biologicheskikh soobshchestv.* (Stability of biological societies.) M.: Nauka,. 352 p;.

57. Chernigovsky V.N. (1981). Habiting and Its Possible Mechanisms. *Izv. AN SSSR. Ser. Biol.*, v.4, p. 485.

58. *Tamar G.* (1972). *Principles of Sensory Physiology.* Springfield, Illinois: Charles & Thomas Publ.

59. Rait R.Kh. (1966). *Nauka o zapakhah (Science about odours).* M.:Mir, 184 p.

60. Romanenko E.V. (1974). *Fizicheskie osnovy bioakustiky (Physical Principles of the Bioacoustics).* M.: Nauka, 178 p.

61. Kan R., Rot G. (1984). *Interaction of Hormones with Receptors.* (Transl. Engl.). M.:Mir, p.10.

62. Marshell E. (1981). *Biophysical Chemistry.* (Transl. Engl.). M.:Mir, v. 1, 2.

63. Soloviev V.N., Firsov A.A. Filov V.A. (1980). *Farmakokinetika* (Pharmacokinetics). M.: Medicina, 423 p.

64. Pradhan S.N., Maikel R.P., Dutta S.N. Ed.(1980). *Pharmacology in Medicine: Principles and Practice.* Bethseda

65. *Gladyshev G.P. and Komarov F.I.* (1996). Hierarchic Thermodynamics and Gerontology. *Vestnik Ross. Med. Acad.*, N 6, p. 31.

66. Shoogo Ueno. Ed. (1991). Biomagnetic Stimulation. New York, London: Plenum Press. *Proceedings of the Intern. Symposium on Biomagnetic Stimulation*, held July 15, 1991. Fukioka, Japan.

67. Flindt von Rainer (1988). *Biologie in Zahlen.* Stuttgart, New York: Gustav Fischer Verlag.

AFTERWORD

Thermodynamics does not consider mechanisms of processes and phenomena, and this is its drawback. The advantage of thermodynamical theory consists in its ability to study the structure transformation of systems in the course of their evolution. The tendency of the evolution of a system on the chemical and supramolecular levels can be revealed regardless of the "surprisingly inconceivable" complexity of the structure of chromatine, organelles, cells, and any other biological objects. The model presented in the book confirms that thermodynamic laws can be applied to any local volume of supramolecular structures and, generally, to any macroscopic volume of the biomass.

There are grounds to suppose that results presented in the book will stimulate further research. There will be further studies aimed at obtaining new evidence that the second law can be applied to the explanation of the evolution of biological and other natural systems. Studying the world requires general methods. I hope that the macrothermodynamic method is one of them.

Differential Equations of Macrothermodynamics.
The Systems and the Processes

As a rule, modern thermodynamics studies simple or complex systems with similar processes taking place in a single fixed time scale. (Usually, the processes are localized in one or several hierarchies.) However, the processes in real systems usually imply complicated transformations involving structures of various hierarchies and relating to different time scales.

Let us consider some differential equations and characteristic functions that are important for macrothermodynamics - thermodynamics *of complex hierarchic natural systems*.

We use the commonly known equation combining the first and the second laws of thermodynamics for complex closed systems where physical-chemical processes take place:

$$TdS \geq dU + pdV - \sum_k X_k dx_k - \sum_k \mu_k dm_k \quad . \tag{1}$$

Here T denotes temperature, S the entropy, U the internal energy, p pressure, V volume, X_k any generalized force except pressure, x_k any generalized coordinate except volume, μ_k chemical potential, m_k the mass of the k-th substance, which can be replaced by the number of moles. The equality sign relates to the case of reversible changes, the inequality describes the irreversible ones. The work performed by the system is negative. (In some literature it is considered to be positive; for instance, see the monographs by K.Denbigh and V.Sychev).

Suppose that in the system under study not only chemical reactions take place but also transformations between the structure elements of other j-th hierarchic levels (i-th partial evolution). Moreover, let the reagents of each i- th evolution condense into the particles of the i-th evolution phase - a structure of higher level of substance organization. In their turn, the latter are reagents of the next, ($i+1$) -th, partial evolution. (The order of the partial evolution, i and of the structure hierarchy, j corresponds to the hierarchy order of the reagents.) Transformation of similar reagents of some hierarchic level (j) into similar reagents of the next hierarchic levels (j+1, j+2,...) can be represented as follows:

$$\ldots mA \underset{t_i}{\overset{\tau_i}{\rightleftharpoons}} A_m \quad\bigg|\quad nA_m \underset{t_{i+1}}{\overset{\tau_{i+1}}{\rightleftharpoons}} (A_m)_n \quad\bigg|\quad p(A_m)_n \underset{t_{i+2}}{\overset{\tau_{i+2}}{\rightleftharpoons}} [(A_m)_n]_p \quad\bigg|\quad r[(A_m)_n]_p \underset{t_{i+3}}{\overset{\tau_{i+3}}{\rightleftharpoons}} \{[(A_m)_n]_p\}_r, \ldots$$

$$\text{\textit{subsystem } j} \quad\quad \text{\textit{subsystem } j+1} \quad\quad \text{\textit{subsystem } j+2} \quad\quad \text{\textit{subsystem } j+3}$$

$$(2)$$

Here n, m, r >>1, and to each j-th structure hierarchy the i-th process of phase-transition type is put into correspondence. Each partial evolution (i, i+1, i+2, ...) takes place in its own time scale t (t_i, t_{i+1}, t_{i+2}, ...), which is comparable with the life-span of the corresponding system or its subsystem (j, $j+1$, $j+2$, ...) of the given structure level (for instance, organells, cells, organisms). At times t the natural subsystems (j) are partly kinetically quasi-closed, i.e., non-stationary. Since a hierarchic system is kinetically closed only with respect to separate components of the flow coming out, it is open as a whole, and therefore, its volume and composition vary. At the same time, in the subsystems of the j-th, (j+1), (j+2) levels local quasi-equilibrium are achieved at small times τ_i, τ_{i+1}, τ_{i+2}, ... These quasi-equilibrium are «drifting» ones, they are achieved at every «moment» during the evolution of the corresponding system. (Different-size arrows in the equalities (2) show that in the process of thermodynamic self-organization - self-assembly we consider namely quasi-equilibrium «drifting to the right»; the large upper arrow shows the direction where the equilibrium is shifted.) Therefore, each partial evolution is a process of self-assembly (i) for the substance of variable composition in an open non-stationary system (j). At any moment of the i-th partial evolution, the subsystem j (for instance, the subsystem of supramolecular structures) can be characterized by the value of Gibbs function of the system formation, $\Delta \overline{\tilde{G}}_{\Sigma_j}$.

One can also use the variation of the specific value of other state function, see Chapter II. The value $\Delta \overline{\tilde{G}}_{\Sigma_j}$ relates to the process (i) of self-assembly (condensation) of the elements of j-th structure level, which are initially in the standard state: for instance, the ideal gas or «ideal solution». As a result of such self-assembly, the products of partial evolution i are formed, their composition varying in time. These products are the reagents of the partial evolution (i+1). One of the partial evolution with the participation of different reagents of the k-th type can be presented as

$$\dots \quad \sum_{k,k'} m_{k'} A_k \underset{t_i}{\overset{\tau_{i,k}}{\rightleftharpoons}} \sum_{k,k'} (A_k)_{m_{k'}} , \quad \dots,$$

$$(3)$$

where $\tau_{i,k}$ do not differ much, and, therefore, the total process of fast relaxation takes place in a single time scale.

We stress that the schemes (2) and (3) describe a real evolution process consisting of partial evolution. These partial evolution are characterized by different mean times of fast relaxation τ_i, τ_{i+1}, ... and go on in the corresponding time scales t_i, t_{i+1}, ... To a certain approximation, these schemes reflect the hierarchy of structures (j) and the corresponding hierarchy of processes (i) in Nature. The evolution of natural biostructures can be characterized as avalanche-condensation structure formation. From the viewpoint of its mechanism, it is opposite to avalanche or branched processes, like electron avalanches accompanying the electric discharge in gases or branched chemical reactions.

Taking into account the schemes (2) and (3), one can write the generalized equation of the first and the second laws in the form:

$$\sum_i T_i^j dS_i^j \geq \sum_i dU_i^j + \sum_i p_i^j dV_i^j - \sum_i \sum_{k_i} X_{k_i}^j dx_{k_i}^j - \sum_i \sum_{k_i} \mu_{k_i}^j dm_{k_i}^j ,$$

$$(4)$$

where the subscript i relating, as before, to the partial evolution and k to the component of the i-th evolution. The upper index j indicates that we consider the behaviour of the j-th system.

Relation (4) introduces the potentials of the components of various partial evolution. It is the most general equation in the thermodynamics of hierarchic structures, for it accounts for any kind of interaction (work), transformations of the substance in the system (variation of its amount in different subsystems), and also the entropy variation in the course of all partial evolution. Note that this relation can be used, to a certain approximation, within any given time scale.

It follows from the relation (4) that the full differentials of the characteristic functions of corresponding characteristic (independent) variables given below can be written for the case of complex system as

$$dU = \sum_i dU_i = \sum_i T_i dS_i - \sum_i p_i dV_i + \sum_i \sum_{k_i} X_{k_i} dx_{k_i} + \sum_i \sum_{k_i} \mu_{k_i} dm_{k_i} \quad , \tag{5}$$

$$dH^* = \sum_i dH_i^* = \sum_i T_i dS_i + \sum_i V_i dp_i - \sum_i \sum_{k_i} x_{k_i} dX_{k_i} + \sum_i \sum_{k_i} \mu_{k_i} dm_{k_i} \quad , \tag{6}$$

$$dF = \sum_i dF_i = -\sum_i S_i dT_i - \sum_i p_i dV_i + \sum_i \sum_{k_i} X_{k_i} dx_{k_i} + \sum_i \sum_{k_i} \mu_{k_i} dm_{k_i} \quad , \tag{7}$$

$$dG^* = \sum_i dG_i^* = -\sum_i S_i dT_i + \sum_i V_i dp_i - \sum_i \sum_{k_i} x_{k_i} dX_{k_i} + \sum_i \sum_{k_i} \mu_{k_i} dm_{k_i} \quad , \tag{8}$$

where the indices j are omitted for convenience and the superscript * means that we study the behaviour of complex systems. The functions H^* and G^*, which do not correspond with their analogues for simple systems, H and G, can be written as $H^* = H - Xx$, $G^* = G - Xx$. The evolution potentials in the equations (5)-(8) are defined as

$$(\partial U / \partial m_{k_i})_{S_i, V_i, x_{k_i}, m \neq m_{k_i}} = (\partial H^* / \partial m_{k_i})_{S_i, p_i, X_{k_i}, m \neq m_{k_i}} =$$
$$= (\partial F / \partial m_{k_i})_{T_i, V_i, x_{k_i}, m \neq m_{k_i}} = (\partial G^* / \partial m_{k_i})_{T_i, p_i, X_{k_i}, m \neq m_{k_i}} = \mu_{k_i} \tag{9}$$

where all corresponding variables of the type m are chosen constant, with the exception for those used as the differentiation variables. Since μ_{k_i} in the equation (9) relates to a complex system, it can be denoted by $\mu_{k_i}^*$.

For a complex hierarchic, say, ecological, system, contributions of different terms in (4)-(9) are determined by the choice of the time scale and the parameters i,k. For instance, for a real system where higher evolution take place involving rather large structures (organisms and so on), the contribution of the terms $T_i dS_i$, $p_i dV_i$ and some others, is extremely small. However, the values $\left|X_{k_i} dx_{k_i}\right|$ and $\left|x_{k_i} dX_{k_i}\right|$ can be rather large in this case. Indeed, in the case of small particles the energy of thermal motion $k_B T >> E_n$. (E_n is the energy of natural magnetic, electric, and gravity fields acting on the particle.) Still, for sufficiently large structures the inverse is valid: $E_n >> k_B T$. Hence, the relative contribution of the corresponding terms is large. While passing from lower evolution to higher ones, one can also observe variations of the relative contributions of the terms characterizing surface, photochemical, magnetic-hydrodynamic, deformational, and other kinds of work.

While studying real processes with fixed time scales, one can simplify considerably the equation (4) by neglecting the small terms. Certainly, the thermodynamics (thermostatic) of *systems* is not interested whether the transition from one state into another one is equilibrium or not. Still it should be noted that there are many real chemical, biological, and other *processes* studied experimentally that go practically in equilibrium regimes. (Such processes are studied by the thermodynamics of equilibrium *processes*.) When fast chemical reactions take place in the case of a stationary flow (the situation is close to the one of the Van't -Hoff box), the actual useful work of the system $(W_x)_{T,p,...}$ approaches the maximal work of the process, i.e., the relation $-(W_x)_{T,p,...} \cong -(W_{rev})_{T,p,...} = -\Delta G^{\cdot}$ holds true. This relation is also satisfied with high accuracy in many natural chemical and biological systems where stationary processes take place. As an example, recall that for the processes in fuel cells the rational efficiency $\eta_R = -(W_x)_{T,p}/-\Delta G^*$ is close to unity. (In the industrial hydrogen-oxygen fuel cells, the value of η_R can reach 0,9 at low currents.) For electrochemical reactions carried out in laboratory conditions, one can easily achieve $\eta_R = 0,99$. Several equilibrium (quasi-equilibrium) chemical transformations in living organisms are described in literature. Thus, we come to the conclusion that many natural processes can be characterized by low irreversibility: to a certain good approximation they can be considered as quasi-static (quasi-equilibrium) ones and investigated by means of the classical thermodynamics of drifting equilibrium. As to the natural chemical and biological systems, in the majority of cases they can be characterized with high accuracy by the corresponding functions of the states (see the Chapter I). Besides, from the principle of structure stabilization and the statements following from macrothermodynamics one can conclude that the rational efficiencies of spontaneous evolutionary *processes* tend to the unity in the course of the evolution $(\eta_R \to 1)$. This conclusion is in good agreement with the general statements of modern ecology.

Certainly, the results presented do not neglect the existence of dissipative structures and the effects of dynamic self-organization in natural systems (I. Prigogine), for instance, when such systems undergo «revolutionary changes» due to the influence of external factors, and their state is far from the equilibrium one.

REFERENCES

1. *Alberty R.A.* (1987). *Physical chemistry*. 7th Ed. New York, etc.: Wiley, 934 p.
2. *Bazarov I.P.* (1983). *Termodinamika.*(Thermodynamics). 3rd Ed. Moscow: Vysshaya shkola, 344 p.
3. *Gladyshev G.P.* (1988). *Termodinamika i Makrokinetika Prirodnykh Ierarkhicheskih Processov* (Thermodynamics and Macrokinetics of Natural Hierarchic Processes). Moscow: Nauka, 287 p.
4. *Gladyshev G.P.* (1995). *Termodinamika of the Ierarkhicheskih Sistem* (Thermodynamics of the Hierarchic Systems). In: *Khimicheskaya Enciklopedia* (Chemical Encyclopedia). M.: v.4, p.535.
5. *Gladyshev G.P., et.al.* (1993). Macrothermodynamics of Biological Systems and Evolution. *J. Biol. Syst.*, v. 1, N 2, p. 115.
6. *Gladyshev G.P.* (1995). Thermodynamic Trends of Biological Evolution. *Izvestia RAN. Seria biol.*, 1995, N 1, p. .5 . (*Biology Bulletin*, v. 22, N 1, p.1. ISSN 1062-3590; N 1, 1995).
7. *Gladyshev G.P.* (1996). Thermodynamic Trends of Biological Evolution. Model and Reality. *Izvestia*

RAN. Seria biol., N 4. (Biology Bulletin ISSN 1062-3590, N 4, p. 315).

8. *Gibbs J.W.* (1928). *The Collected Works of J. Willard Gibbs. Thermodynamics.* V.1. New York: Longmans, Green and Co. (Russian transl.: Moscow: Mir, 1950, 444 p.).

9. *Guggenheim E.A.* (1977). *Thermodynamics: An Advanced Treatment for Chemists and Physicists.* 6th Ed. Amsterdam etc.: North Holland, 390 p.

10. *Denbigh K.G.* (1971). *The Principles of Chemical Equilibrium.* 3rd Ed Cambridge: Cambridge Univ. Press, 491 p.

11. *Kubo R.* (1968). *Thermodynamics.* Amsterdam: North-Holland Publ. Co. (Russian transl.: Moscow: Mir, 1970, 304 p.).

13. *Ovsyannikov A.A.* (1995). Samoorganizatsiya. (Self-organization). In: *Khimicheskaya Enciklopedia* (Chemical Encyclopedia). M.: v.4, p. 574.

14. Relaksatsiya. (Relaxation) (1995). In: *Khimicheskaya Ensiklopedia* (Chemical Encyclopedia). M.: v.4, p. 462.

15. *Sychev V.V.* (1986). *Slozhnye termodinamicheskie sistemy* (Complex thermodynamic systems). M.: Energoatomizdat, 208 p. (Translated into English: M.: Mir, 1985).

16. *Haywood R.W.* (1980). *Equilibrium Thermodynamics.* Chichester, New York, Brisbane, Toronto: A Wiley Interscience Publ. (Russian transl.: Moscow: Mir, 1983, 491 p.).

SOME NOTIONS AND TERMS USED IN THE BOOK[*]

Complex thermodynamic system - a thermodynamic system where some work aside from extension work is performed (V.Sychev, see reference to Chapter I).

Current quasi-equilibrium - instant equilibrium achieved by a system as a result of fast relaxation. CQE always exists in a quasi-equilibrium, quasi-static process. One can understand CQE as a stationary (time-independent) or quasi-stationary nonequilibrium state of an open system, stable with respect to small perturbations.

Dynamic self-organisation - (or simply self-organisation, in the terminology of I.Prigogine) - a process consisting in the arising, reproduction, or improvement of the organisation of a dynamic system that is far from equilibrium.

Evolution potentials, μ_{k_i}, - specific thermodynamic potentials that determine the changes in thermodynamic potentials (functions) resulting from the variations of the number of particles of certain nature (components of the i-th partial evolutions) of the type k. The EP are analogues of the chemical potential and are used in hierarchic thermodynamics for the description of systems with variable number of particles. The EP can be used in macrothermodynamics for describing the

---

[*] Appendix 2 contains more accurate definitions of notions and terms used in the monograph. (Some of the formulations have been given by the author in his earlier publications.)

behaviour and evolution of open complex hierarchic non-stationary systems. For example, $\mu_{k_i}^{j}$ is the potential of the k-th component, i-th constituent part of the process (partial evolution) taking place in the system j. This notation, used in hierarchic thermodynamics, allows to present the chemical potential of the k-th chemical component in the j-th ideal system as $\mu_{k_{ch}}^{j}$.

Gibbs function (G), or the fundamental Gibbs function, the Gibbs potential, the Gibbs free energy, free energy at constant pressure, free enthalpy, isobaric potential, isobaric-isothermic potential, thermodynamic potential. The GF of a physical-chemical system can be divided into the chemical (molecular) component (ch) characterising the chemical structure of matter, and the supramolecular (im) component characterising the supramolecular structure of matter: $G = G^{ch} + G^{im}$. (These structures are formed by chemical and supramolecular bonds, respectively.) Correspondingly, the GF of the formation of condensed matter is represented in the form $\Delta G = \Delta G^{ch} + \Delta G^{im}$.

This distinguishing is, in principle, similar to dividing the observed variable ΔG_{obs} into the ideal component, ΔG_{ideal}, and the excess component ΔG^{E}, which is the Gibbs function of mixing: $\Delta G^{E} = \Delta G_{obs} - \Delta G_{ideal}$. The value of ΔG_{ideal}, for solutions is calculated using the formulas for ideal systems obeying the Raoult`s law for any composition of mixture.

In physical chemistry, one often uses the following values: the Gibbs function ΔG_{f} of the formation of a compound from the elements and the standard Gibbs function ΔG_f^0 of the formation of a compound from the elements. In hierarchic thermodynamics, if necessary, these values can be denoted as ΔG_f^{ch} and $\Delta G_f^{ch\,0}$. According to this notation, the Gibbs function of the formation of the superstructure from chemical compounds, or from a separate compound, can be expressed as ΔG_f^{im}, and the standard function as $\Delta G_f^{im\,0}$.

Hierarchic thermodynamic system - thermodynamic system consisting of hierarchic subsystems that are related to each other by structure and may be other subordination and by the transitions from lower levels to higher ones. These subsystems should be also separated in space and (or) with respect to the time needed for the relaxation to equilibrium.

Hierarchic thermodynamics (macrothermodynamics, or structure thermodynamics) studies complex heterogeneous chemical and biological systems, first of all, open systems that exchange matter and energy with the environment. According to the approach of HT, such a system should be represented as a set of subordinate subsystems related hierarchically by their positions in space (structural, or spatial hierarchy) and (or) in time (time hierarchy).

The central notion of HT is the notion of partial evolution (the *i*-th process), i.e., aggregation of the k_i-th components of the system participating in the process *i* on the level *j*. For example, the negligible non-equilibrium self-assembly process taking place under the melting point of matter can be considered as aggregation of molecules and macromolecules resulting in the formation of supramolecular structures.

Kinetically quasi-closed system - a thermodynamic system open at relatively large times, through which a flow of substance passes and in which, due to thermodynamic factors, comparatively stable substances (for instance, supramolecular structures) are accumulated. This substance accumulation is considered as partial quasi-closeness of the system with respect to some components of the out-coming flow of matter. The system is non-stationary at its times of existence. Kineticaly quasi-closeness of the corresponding subsystems in biological objects is ensured by the hierarchic sequence of thermostats. The existence of this sequence is caused by unidirected series of life (relaxation) - times for structures of different hierarchies.

Life-span (life-time) - the average existence time for a particle or a system (t or $\bar{t}$). For non-stable radioactive isotopes or chemical compounds dissociating via first-order reactions, the life-span is defined as $t = T_{1/2}/\ln 2 = 1/k$, k - being the rate constant, $T_{1/2}$ - the half of the life-time. The LS of monomer molecules fed into a polymerization system is determined as the average time spent by a molecule in unbonded (monomer) state. Analogously, one defines the LS of free radicals generated, for instance, by light. The LS of living objects is defined as the life duration of a biosystem, some biostructure (for instance, an organelle, a cell, or an organism), which is defined by the moments of birth and death.

Macrothermodynamics (or hierarchic thermodynamics) studies all sorts of heterogeneous systems (simple and complex) using the methods of thermostatics and non-equilibrium thermodynamics. For example, M of open hierarchic non-stationary systems describes the thermodynamic behaviour of natural systems, for instance, biological ones. The part "macro" in the term "macrothermodynamics" is used to stress that this branch of science studies heterogeneous (polyhierarchic) macroobjects. At the same time, thermodynamics of any systems or processes describes the behaviour of systems only on the macroscopic level. From this viewpoint, the part "macro" in the term "macrothermodynamics" does not possess any special physical sense.

Ontogenesis - individual development of a living thing, all sequence of its transformations from birth to the end of life.

Partial equilibrium - equilibrium of a physical system with respect to a single or several parameters x_i. A system in PR with respect to the parameter x_i, can stay non-equilibrium with respect to other parameters δ_k. For instance, real quasi-equilibrium in hierarchic thermodynamics is a PE established after some i-th partial evolution is accomplished in the given system.

Partial evolution (in hierarchic thermodynamics) - the process of self-assembly (thermodynamic self-organisation) of structures relating to the i-th hierarchy resulting in the formation of structures of a higher hierarchy - (j+1). For instance, the self-assembly of molecules leading to the formation of supramolecular structures can be considered as partial molecular evolution.

Philogenesis - historical development of the world of living organisms both as a whole and in separate taxonomic groups: kingdoms, types, classes, ordos, families, genuses, species.

Population - a group of organisms belonging to a single species, possessing a common genetic fund occupying a definite territory, and, as a rule, more or less isolated from other similar groups.

From the zoological viewpoint, Homo sapiens is a species widely (though inhomogeneously) spread over Earth including numerous populations.

The principle of the stability of a chemical substance is a set of qualititaive regularities according to which the relatively low chemical (*ch*) thermodynamic stability of a compound in a state of ideal gas or solution (ΔG_f^{ch}) causes relatively high supra-molecular (intermolecular, *im*) thermodynamic stability of condensed phases formed by that compound (ΔG^{im}). Conversely, the higher the chemical thermodynamic stability of a substance, the lower its supra-molecular thermodynaic stability in a condensed state. This regularity can be expressed through values of $\Delta \acute{I}_{comb}$, ΔH_{form} and other functions.

The principle was applied by the author to various hierarchies as part of the theory of the evolution of life. It is in agreement with the principle of structural stabilization.

The principle of the stability of a chemical substance can be applied with a number of qualifications to multi-component systems (Figures 4 and 7) and reflects the tendency of atoms of different elements to condense around other atoms through chemical and non-chemical (intermolecular, *im*) ties. The above regularities can be identified for isoatomic substances, several homologous series, lyophilic, lyophobic and biophilic chemical compounds by building the folowing dependencies:

$$\Delta G_j^{ch} + \alpha_1 \Delta G_j^{im} = const \ ,$$

$$\Delta H_j^{ch} + \alpha_2 \Delta H_j^{im} = const \ ,$$

where α_1, α_2 - coeffcients and the subscript j relates to the substance.

The available data demonstrate that the above relationships cover most simple substances soluble in water.

Precise mathematical formulation of the principle may be impeded, because the measurement of the absolute values of thermodynamic functions is not possible.

Principle of structure stabilisation - the principle stating that in hierarchic systems each higher-level partial evolution or, in other words, higher i-th partial process, which is a component of the general evolution - the process of hierarchic structure formation - stabilises, due to the aggregation, the products of the lower partial evolution, or $(i-1)$-th partial process. For instance, in a non-stationary open system where various chemical reactions and other processes take place, structure stabilization "chooses" supramolecular structures that are most stable thermodynamically.

Thermodynamic equilibrium - the state of a thermodynamic system with parameters constant in time. Any system isolated from the environment would spontaneously tend to the TE. As soon as the system reaches TE, all irreversible changes are stopped in it. There exist several conditions of TE relating to the establishment of mechanical, thermal, chemical, social and other types of equilibria in a system. Mechanical equilibrium implies that any macroscopic movements of the parts of the system are prohibited, though the system can move or rotate as a whole. The thermal and chemical equilibria in a system are determined by the condition that temperature and chemical potentials are constant in the volume of the system (or its local microvolumes that can be considered as isolated). Hierarchic thermodynamics also considers other types of equilibrium, which are characterised by constant population, sociological, or other potentials. Like chemical potential, these potentials are defined for the components of the corresponding hierarchies of structures and consideration levels. The conditions for the TE to be stable are obtained from the second law of thermodynamics. It states that the thermodynamic potential of a system is minimal at the TE, while the entropy in this case is maximal (for the corresponding independent variables, i.e., charac-teristic variables).

Quasi-closed system - a thermodynamic system, open at comparatively large times, which can be considered, to a certain approximation, as closed, due to the fast relaxation to local equilibria (such as supramolecular equilibrium) and (or) due to the partial accumulation of matter coming into the system. In accordance with this general definition, it is worth distinguishing between thermodynamical and kinetical quasi-closeness of a system.

Quasi-static process, or equilibrium process, - an infinitely slow transition of a thermodynamic system from one equilibrium state into another, such that at every time moment the state of the system is infinitely close to the equilibrium state. During a QSP, the system reaches equilibrium much faster (almost instantaneously) than its physical parameters vary. A QSP is not necessarily a reversible process.

Quasi-stationary state in physics - the same as metastable state. A QSS in chemistry and biology is the state of a system involved into a reaction; this state, to a certain approximation, is characterised by constant concentrations of the intermediate products. The concept of QSS can be valid for the description of systems in certain time scales but invalid in other time scales. For instance, a cell, which is a biological system, is not a stationary system at its life-time t. However, at smaller times $\tau \ll t$ its behaviour can be considered as stationary; more precisely, quasi-stationary. At the moment of death the cell passes to the state of partial equilibrium (real quasi-equilibrium).

Real quasi-equilibrium - in hierarchic biological thermodynamics (macrothermodynamics), the state of an open system (considered at the given hierarchic level) that is achieved at the moment of death due to the accomplishment of the substance exchange with the environment in the given time scale.

Relaxation - the process leading to the establishment of thermodynamic equilibrium in macroscopic thermodynamic systems. It should be taken into account that the equilibrium state can be determined by a large number of parameters, and the processes of achieving equilibria with respect to different parameters can go in different ways and at different rates. R is quantitatively characterised by the relaxation time.

Simple thermodynamic system - a thermodynamic system where no work or only extension work is performed (V.Sychev, see reference to Chapter I).

Society - any group of organisms belonging to different species coexisting at some territory and interacting through trophic and spatial relations.

Sometimes one specifies S. of plants (phytocenos) and S. of animals (zoocenos). A S. is a system of definite hierarchic level of living matter organisation. Elements of S. are populations of different species. The S. itself is an element of an ecosystem (or biogeocenos).

With certain constraints, one can also use the term "society of people" as a species including numerous populations.

Stationary state in physics - a state of a physical system where parameters essential for its description do not vary in time. For instance, if the velocity of a flow of fluid is constant at every point, then the state of the flow is stationary. In chemistry, the state of a chemical system involved into a reaction is called SS if the intermediate products of the reaction have constant concentration. In a system with flow (a reactor), concentrations of the components are constant for the SS. In the limiting case, when the flows of matter in an open system tend to zero, the system reaches equilibrium state.

Supramolecular composition of a system - composition of a system consisting of supramolecular components of different nature. For instance, the supramolecular components of a biological membrane, which is a heterogeneous system, are supramolecular formations (aggregates) of proteins, polysugars, lipids, etc.

Synergetics - a frontier branch of science revealing general tendencies in the processes of formation, stability, and destruction of ordered temporal and spatial structures in complex systems of various nature, which are far from equilibrium. The models of synergetics are models of non-equilibrium systems in the presence of fluctuations.

Thermodynamic self-organisation (self-assembly) - spontaneous ordered joining of the structures of *i*-th hierarchy into structures of (*i*+1)-th hierarchy. The process of self-assembly (or partial evolution) is a weakly non-equilibrium process similar to phase transition. For instance, formation of supramolecular structures from molecules in a cell can be considered as a phase transition from over-cooled state. Thermodynamic self-organisation is observed in systems close to equilibrium.

TS on the physical-chemical level leads to the association of molecules, macromolecules, or their aggregates. At this process, intermolecular interactions induce changes in the internal structure of molecules, which lead to the appearance of new conformations.

Thermodynamically quasi-closed system - a thermodynamic system open at relatively large times, which can be considered as closed at small times (due to the negligibly small rate of the matter exchange with the environment). For instance, macromolecules in the biotissues of living organisms under physiological conditions form conformations corresponding to the minimum of the Gibbs function; this indicates that the system "macromolecule - environment" is closed.

Thermostat in the thermodynamics of complex systems - a part of the total system that can be considered as surrounding or environment. The T. imposes certain conditions on the system under study, which is a subsystem of the total system. These conditions can be constant temperature, pressure, chemical potentials or any other potentials (for instance, sociological ones), etc. To avoid the confusion between the definition of T. as a thermal reservoir and the definition given above, it is worth indicating what sense of the term is meant (see R.Kubo, reference to Chapter I).

Time-scale of a process (for instance, rolling up of molecules and formation of globular structures, ontogenesis, philogenesis, etc.) - time interval that is approximately equal to the duration of the process (phenomenon) under study. For instance, the ageing (ontogenesis) of human can be studied in the time scale of about 100 years.

Unidirected series of relaxation times (life-times, or existence times) in hierarchic thermodynamics - a sequence of relaxation times or life-times for structures of different hierarchic levels ordered into a series such that its neighbouring terms are connected by unidirected strong inequalities. USRT reflects one of the fundamental natural laws.

WHAT DO THERMODYNAMICS AND DYNAMICS STUDY?

Equilibrium thermodynamics - thermostatics

Systems	**Processes**
Thermodynamics of systems	Thermodynamics of processes

The state of the studied objects is determined by state functions

Example	*Example*
The system "fluid - vapour" in equilibrium	Type I phase transition

Non-equilibrium thermodynamics

Systems		**Processes**	
Thermodynamics	Dynamics	Thermodynamics	Dynamics
Systems close to equilibrium	***Systems far from equilibrium***	***Weakly non-equilibrium processes***	***Strongly non-equilibrium processes***
A system can be characterised by state functions	*A system can be characterised by dynamic functions*	*A process can be studied via the variation of the state functions*	*A process can be studied via the variation of the dynamic functions of the system*

Examples

Thermodynamically self-organised-self-assembled macromolecular system	A system of excited particles	Thermodynamical self-organisation - self-assembly of molecules	Dynamic self-organisation (self-organisation) resulting in the formation of dissipative structures

In this paper a special attention is paid to the thermodynamics (thermostatics) of systems - biostructures that can be described at any moment by state functions (in particular, the Gibbs function) having real physical sense.

INDEX

T

U